Springer Finance

Springer-Verlag Berlin Heidelberg GmbH

Springer Finance

Springer Finance is a programme of books aimed at students, academics and practitioners working on increasingly technical approaches to the analysis of financial markets. It aims to cover a variety of topics, not only mathematical finance but foreign exchanges, term structure, risk management, portfolio theory, equity derivatives, and financial economics.

Credit Risk Valuation: Methods, Models, and Application
M. Ammann
ISBN 3-540-67805-0 (2001)

Risk-Neutral Valuation: Pricing and Hedging of Finance Derivatives
N.H. Bingham and R. Kiesel
ISBN 1-85233-001-5 (1998)

Credit Risk: Modeling, Valuation and Hedging
T.R. Bielecki and M. Rutkowski
ISBN 3-540-67593-0 (2001)

Interest Rate Models: Theory and Practice
D. Brigo and F. Mercurio
ISBN 3-540-41772-9 (2001)

Visual Explorations in Finance with Self-Organizing Maps
G. Deboeck and T. Kohonen (Editors)
ISBN 3-540-76266-3 (1998)

Mathematics of Financial Markets
R. J. Elliott and P. E. Kopp
ISBN 0-387-98533-0 (1999)

Mathematical Finance - Bachelier Congress 2000
H. Geman, D. Madan, S.R. Pliska and T. Vorst (Editors)
ISBN 3-540-67781-X (2001)

Mathematical Models of Financial Derivatives
Y.-K. Kwok
ISBN 981-3083-25-5 (1998)

Efficient Methods for Valuing Interest Rate Derivatives
A. Pelsser
ISBN 1-85233-304-9 (2000)

Exponential Functionals of Brownian Motion and Related Processes
M. Yor
ISBN 3-540-65943-9 (2001)

Interest Rate Management
R. Zagst
ISBN 3-540-67594-9 (2002)

Robert Buff

Uncertain Volatility Models- Theory and Application

Springer

Robert Buff
Goldman Sachs & Co.
85 Broad Street
New York, NY 10004
USA
e-mail: uvm@robertbuff.com

Mathematics Subject Classification (2000): 91B24, 91B26, 91B28, 65N06
JEL Classification: C63

Cataloging-in-Publication Data applied for

Die Deutsche Bibliothek - CIP Einheitsaufnahme

Buff, Robert:
Uncertain volatility models: theory and application/Robert Buff.-Berlin; Heidelberg; New York; Barcelona; Hongkong; London; Mailand; Paris; Tokio: Springer, 2002
(Springer finance)
Additional material to this book can be downloaded from http://extras.springer.com.

ISBN 978-3-540-42657-8   ISBN 978-3-642-56323-2 (eBook)
DOI 10.1007/978-3-642-56323-2

This work is subject to copyright. All rights are reserved, whether the whole or part of the material is concerned, specifically the rights of translation, reprinting, reuse of illustrations, recitation, broadcasting, reproduction on microfilm or in any other way, and storage in data banks. Duplication of this publication or parts thereof is permitted only under the provisions of the German Copyright Law of September 9, 1965, in its current version, and permission for use must always be obtained from Springer-Verlag. Violations are liable for prosecution under the German Copyright Law.

http://www.springer.de

© Springer-Verlag Berlin Heidelberg 2002

The use of general descriptive names, registered names, trademarks etc. in this publication does not imply, even in the absence of a specific statement, that such names are exempt from the relevant protective laws and regulations and therefore free for general use.

Please note: The software is protected by copyright. The publisher and the author accept no legal responsibility for any damage caused by improper use of the instructions and programs contained in this book and the CD-ROM. Although the software has been tested with extreme care, errors in the software cannot be excluded. The software may only be used under the terms of the GNU General Public License (GPL), Version 2 or later, as stated in the instructions on the CD-ROM.

Cover design: *design & production*, Heidelberg
Typesetting by the author using a Springer LaTeX macro package
Printed on acid-free paper   SPIN 10850758   41/3142db-5 4 3 2 1 0

For Chao-Mei

Preface

Many introductory books on mathematical finance also outline some computer algorithms. My goal is to contribute a closer look at algorithmic issues that arise from complex forms of the underlying pricing models—issues many practitioners need to solve sooner or later in their careers.

This book takes such a close look at uncertain volatility models, an extension of Black-Scholes theory. It discusses applications to exotic option portfolios with barriers and early exercise features. It describes an object-oriented C++ solution, included in source code on the accompanying CD.

Practitioners and students who need to build analytic software libraries may benefit from reading this book and studying the software. The book focuses on a family of mathematical models, while in the field one encounters greater variation in instrument properties. In both cases mathematical and financial knowledge must be complemented by good programming skills to produce the best system. Analytic software needs design—a central message of the later chapters of this book.

This book has come out of my Ph.D. thesis. I am very grateful to my academic advisor, Marco Avellaneda of New York University, who taught me mathematical finance and uncertain volatility. Computational finance became exciting for me because Marco encouraged an algorithmic approach to uncertain volatility.

I thank Afshin Bayrooti, Vladimir Finkelstein, and Antonio Parás for giving valuable feedback. Antonio is the co-inventor of the original uncertain volatility model, λ-UVM.

Richard Holmes has found a crucial bug in an early implementation of the software.

I thank my editor Catriona Byrne, Daniela Brandt and Susanne Denskus from Springer Verlag for the production of this book.

Chao-Mei Wei has seen the final stages of my research, my entry into Wall Street, and the realization of this book while she was writing hers. She will always be the motivation behind my work.

New York, *Robert Buff*
December 2001

Table of Contents

1 **Introduction** 1
   1.1 Uncertain Volatility Scenarios and Exotic Options 2
   1.2 Volatility Shock Scenarios 3
   1.3 Object-Oriented Implementation 4
   1.4 User Interfaces: Scripting and Mathematica 5
   1.5 How to Best Read This Book 6

Part I Computational Finance: Theory

2 **Notation and Basic Definitions** 11
   2.1 Linear Algebra 11
   2.2 Probability and Stochastic Processes 11
   2.3 Partial Portfolios and Positions 12
   2.4 Accents, Superscript, Subscript 13

3 **Continuous Time Finance** 15
   3.1 Deterministic Volatility 15
      3.1.1 One-Factor Black-Scholes Analysis 15
      3.1.2 Hedging with Black-Scholes 17
      3.1.3 Interest Rate Models 19
   3.2 Stochastic Volatility 24
      3.2.1 Tradable and Nontradable Factors 24
      3.2.2 Some Concrete One-Dimensional Models 25
   3.3 Model Calibration 30
      3.3.1 Parametric Methods 30
      3.3.2 Non-Parametric Methods 31

4 **Scenario-Based Evaluation and Uncertainty** 33
   4.1 Preliminaries 33
   4.2 The Worst-Case Volatility Scenario 35
      4.2.1 Worst-Case Pricing 36
      4.2.2 The Optimal Hedge Portfolio 38
      4.2.3 Calibration to the Worst Case 39
   4.3 Minimum-Entropy Calibration 41

X   Table of Contents

      4.4   Scenarios and Nonlinearity 43

Part II Algorithms for Uncertain Volatility Models

5 A Lattice Framework 47
    5.1 Multi-Lattice Dynamic Programming 48
        5.1.1 Data Structures 49
        5.1.2 Dataflow for Explicit Methods 50
        5.1.3 Dataflow for Mixed Explicit/Implicit Methods 51
    5.2 Numerical Issues 51

6 Algorithms for Vanilla Options 57

7 Algorithms for Barrier Options 61
    7.1 The Hierarchy of PDE's 63
        7.1.1 Construction 63
        7.1.2 Complexity 66
    7.2 Empirical Results 71
        7.2.1 Numerical Convergence 71
        7.2.2 Introducing Uncertainty 73

8 Algorithms for American Options 77
    8.1 Early Exercise Combinations 78
        8.1.1 Long and Short Positions 78
        8.1.2 Best Worst-Case Evaluation Formalized 81
    8.2 Speedup Techniques 90
        8.2.1 Maintaining the Corridor of Uncertainty 93
        8.2.2 Collapsing the Corridor of Uncertainty 98
        8.2.3 Miscellaneous Issues 105
    8.3 Empirical Results 110
        8.3.1 Computational Complexity 110
        8.3.2 Stress Tests 115
    8.4 American Options and Calibration 122

9 Exotic Volatility Scenarios 123
    9.1 Volatility Shocks for Portfolios of Vanilla Options ... 123
        9.1.1 Worst-Case Volatility Shocks 125
        9.1.2 Empirical Results 133
    9.2 Volatility Shocks and Exotic Options 138

Part III Object-Oriented Implementation

10 The Architecture of Mtg 143

11	**The Class Hierarchy of MtgLib—External**	145
	11.1 Instruments	146
	11.2 Portfolios	151
	11.3 Models	152
	11.4 Model Coefficients	158
	11.4.1 The Base Class `tTermStruct`	159
	11.4.2 Classes Derived from `tTermStruct`	162
	11.4.3 Classes with `tTermStruct` Components	162
	11.5 Scenarios	166
	11.5.1 The Base Class `tScenario`	167
	11.5.2 Classes Derived from `tScenario`	170
	11.6 Numerical Methods	171
	11.6.1 Lattice Templates and Instances	172
	11.6.2 Finite Difference Solvers	179
	11.7 Evaluators	182
12	**The Class Hierarchy of MtgLib—Internal**	185
	12.1 Compute Engines	185
	12.1.1 The Abstract Class `tEngine`	186
	12.1.2 The Abstract Class `tFDEngine`	188
	12.1.3 The Abstract Class `tOFEngine`	190
	12.1.4 The Concrete Class `tGeoEngine`	193
	12.1.5 The Concrete Class `tShockEngine`	193
	12.2 Other Groups of Classes	194
13	**Extensions for Monte-Carlo Pricing and Calibration**	195
A	**The Network Application MtgClt/MtgSvr**	197
B	**The Scripting Language MtgScript**	203
	B.1 Factor Objects	204
	B.2 Claim and Portfolio Objects	205
	B.3 Model, Drift and Volatility Objects	208
	B.4 Lattice and Path Space Objects	212
	B.5 Bootstrapping, Curve and Image Objects	213
	B.6 Scenario and Optimizer Objects	216
	B.7 Evaluation Objects and Examples	217
C	**Mathematica Extensions**	227
	C.1 The Syntax of Object Expressions	227
	C.2 Turning Scripts into Functions	230
	C.3 Profiling and Diagrams	231
References		233
Index		237

1 Introduction

There is a growing collection of literature on standard derivative pricing models that, to a more or lesser degree, contain recipes on how these models are implemented on a computer. In many cases recipes are explained on a very high level or focus on only one link in the chain of components that make up a software system for derivatives pricing.

In this book we travel the entire road from innovative mathematical finance to a working software system, which we call Mtg (for "Martingale"). We discuss the theory, numerical methods, combinatorial algorithms, software architecture and user interface. The software, contained on the accompanying CD for most Windows platforms, can be used through a command-line interface, as a C++ library, through Mathematica, and over the Web.

The software we describe in this book solves the worst-case pricing problem for portfolios of barrier or American options on equity or foreign exchange under uncertain volatility assumptions. Uncertain volatility is computationally interesting because it makes pricing non-linear: the price of two options computed separately may be different from the price of the combined portfolio. We have also added a new type of uncertain volatility scenario in which we allow a fixed number of periods of limited duration with volatility shocks of high amplitude.

The software on the accompanying CD consists of 81500 lines of C++ code and 11500 lines of Java code. Large programs are best developed through modularization. The unit of modularization under the object-oriented approach is the class—an abstract data type that inherits properties from parent classes or defers the instantiation of properties to child classes. Our code contains classes and class hierarchies for entities that have direct financial significance (instruments, portfolios, models, scenarios), classes that support certain mathematical methods (lattices, finite difference solvers, path spaces, optimizers), classes that control the evaluation loop (compute engines and evaluators), scripting classes (parser, scanner, script sources, expressions), system classes (sockets, pipes, services), and others.

1.1 Uncertain Volatility Scenarios and Exotic Options

It is widely accepted that the assumption of constant volatility in financial models (such as the original Black-Scholes model) are incompatible with derivatives prices observed in the market. This problem has several principle solutions: volatility can be made a deterministic function of time and the underlying asset price; or it can be made stochastic, introducing one additional source of randomness. To find this deterministic function (or "volatility surface") or random process, however, poses difficulties in the framework of arbitrage pricing theory since volatility is not directly traded.

Strongly related to finding the right way to model volatility is the problem to measure the exposure of options portfolios to volatility risk; how does the model value of the portfolio change if the volatility assumptions turn out to be false?

We approach these problems by using the uncertain volatility model developed by Avellaneda and Parás as a starting point (see [3]). Uncertain volatility models select a concrete volatility surface among a candidate set of volatility surfaces, and answer the sensitivity question by computing an upper bound for the value of the portfolio under any candidate volatility. (By inverting the position, a lower bound can be computed as well.) This is achieved by choosing the local volatility $\sigma(S_t, t)$ among two extremal values $\sigma_{\min}$ and $\sigma_{\max}$ such that the value of the portfolio is maximized locally.

Uncertain volatility scenarios generalize this approach: given a model that exhibits uncertainty in some of its coefficients (the volatility, in particular), instantiate those uncertain coefficients such that some objective is fulfilled. This objective is called a *scenario*.

The original uncertain volatility model in [3] is a worst-case scenario for the sell-side. By maximizing the portfolio value and charging accordingly, sellers are guaranteed coverage against adverse market behavior if the realized volatility belongs to the candidate set. Worst-case prices are nonlinear, due to diversification of volatility risk. Worst-case evaluation is based on a nonlinear Hamilton-Jacobi-Bellman equation that generalizes Black-Scholes by adjusting the local volatility, or conditional variance, based on the local gamma.

This worst-case volatility scenario has been applied to portfolios of vanilla options, for which the Hamilton-Jacobi-Bellman equation is straightforward to implement on a computer. An extension that hedges a portfolio of vanilla options with liquidly traded market benchmarks is presented in [4].

The computational overhead, however, grows quite dramatically once path-dependent options such as barrier or American options are added to the portfolio. The worst-case volatility scenario from today's perspective of a portfolio containing an American option, for instance, depends on whether the option is exercised today or not (for simplicity, assume the option can be exercised only at finitely many times). A worst-case pricing algorithm must compare

- the worst-case price of the portfolio under the assumption that the American option is exercised tomorrow at the earliest;
- the worst-case price of the portfolio minus the American option, plus the cashflow received or paid immediately due to early exercise.

The pricing algorithm selects the early exercise strategy that works best under the worst-case assumption. As the number of American options in the portfolio increases, the number of different early exercise strategies that must be tested increases exponentially, as nonlinearity forces the pricer to consider all relevant combinations. This leads to a hierarchy of interdependent PDE's, each solving a Hamilton-Jacobi-Bellman problem.

In this book, we solve the pricing problem for portfolios containing barrier and American options under worst-case volatility scenarios. For barrier options, the computational complexity can be determined beforehand and is always $O(n^2)$, n being the number of barrier options in the portfolio. For American options,

- the early exercise boundaries are not known a priori: each PDE describes a free boundary problem, the boundary value being selected locally from a hierarchy of subordinate PDE's (numerical aspect);
- the pricer must distinguish between long and short positions, as agents can use their long positions to counter somewhat the worst-case early exercise strategies ascribed to the investors with whom they have established their short positions. This gives rise to the notion of best worst-case scenario (combinatorial aspect).

Potentially, up to $O(2^n)$ early exercise combinations need to be considered (n being the number of American options in the portfolio). This, of course, is unacceptably expensive. We have developed algorithms that reduce the number of combinations tested locally, but remain correct in the sense that, locally, the best worst-case scenario is always found. We also present a heuristic which reduces the compute time further, but is no longer guaranteed to be correct.

1.2 Volatility Shock Scenarios

Worst-case volatility scenarios limit the candidate set to volatilities that oscillate between two extremal bounds. The resulting spread between the worst-case values for the original and inverted position is often unacceptably large. To narrow the extremal bounds $\sigma_{\min}$ and $\sigma_{\max}$ is a possible solution, but also makes it less likely that the volatility realized later indeed observes those bounds. To narrow the extremal bounds selectively in some places, and leave them unmodified (or even widened) in others seems a plausible alternative, allowing for periods of relative calm and periods of *volatility shocks* with high amplitude.

Where on the time axis should those periods of high volatility fluctuation be located? If market events that influence volatility cannot be foreseen, the exact location of volatility shocks is difficult to determine. The worst-case paradigm comes to the rescue: it is the pricing algorithm's task to locate volatility shock periods where they cause the most damage, in a path-dependent way. Thus, the portfolio value is not only maximized over the local volatility, but also over the location of volatility shock periods.

An example helps to clarify. Suppose the volatility is estimated at 15%. There exists very likely, we assume, one short period of 3 days during which the volatility may vary between 15 and 100%. Given some portfolio, what is its worst-case value under the assumption that the 3-day volatility shock period can start anytime? Its start date may even be path-dependent: it may start earlier if the price of the underlying goes up, and later if it goes down.

Volatility shock scenarios can be solved with dynamic programming. We have developed algorithms that solve volatility shock scenarios for portfolios of vanilla, barrier and American options. The number f of volatility shock periods is not limited. The computational overhead is linear in f: if each volatility shock period lasts exactly one day, for example, then tests show that the computation takes about $3f$ as much time as a regular worst-case scenario.

Volatility shock scenarios are a useful new member in the arsenal of tools that manage volatility risk.

1.3 Object-Oriented Implementation

The second part of the book describes the architecture of the software system Mtg for uncertain volatility models. In the book we emphasize structure over implementation details. The source code for Mtg is contained on the accompanying CD.

Mtg consists of modules MtgLib, MtgClt, MtgSvr, and MtgMath. MtgLib contains the vast majority of the code. MtgSvr and MtgMath are wrappers around the C++ class library MtgLib, offering different ways to access its features. MtgClt is a Java front-end to MtgSvr.

MtgLib contains object-oriented code that solves the worst-case volatility and volatility shock scenarios for vanilla, barrier and American options. Its higher-level combinatorial classes are geared towards multi-factor models on lattices. Its numerical classes for finite difference solutions (explicit and mixed implicit/explicit) accept any one-factor model. MtgLib strictly adheres to the scenario concept.

Figure 1.1 shows how successive refinement leads from a general view on evaluation to a concrete lattice-based method supporting a one-factor Black-Scholes model. At each level in the hierarchy, alternative approaches can be spawned off.

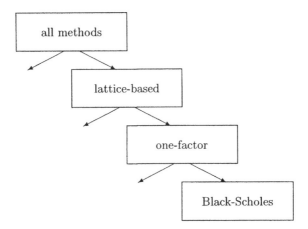

Fig. 1.1. Progressing from a general view on evaluation to the concrete method for the concrete model. The boxes correspond to classes tEngine, tFDEngine, tOFEngine and tGeoEngine in MtgLib

A similar hierarchy can be drawn for scenarios: scenarios in general are refined to worst-case volatility scenarios, which in turn are extended to volatility shock-scenarios. The choice of the scenario is orthogonal to the choice of the method of evaluation. At the deepest level, Black-Scholes may be evaluated under either scenario.

1.4 User Interfaces: Scripting and Mathematica

MtgLib parses portfolio and scenario specifications in its own scripting language, "MtgScript." Users do not have to write in MtgScript themselves, but can use the front-ends MtgClt/MtgSvr or MtgMath.

MtgMath is a wrapper that translates Mathematica expressions into MtgScript. Fig. 1.2 shows an example Mathematica notebook that analyzes a butterfly spread.

The other wrapper program, MtgSvr, is a general-purpose server that accepts requests in the scripting language MtgScript and returns the result in text format. MtgClt is a Java front-end to MtgSvr. It connects to MtgSvr through the TCP protocol. Through MtgSvr, MtgClt handles all cases of exotic options (barrier, American) and volatility scenarios (worst-case, volatility shock) discussed in this book. MtgSvr uses lattice-based numerical methods, but also supports Monte Carlo simulation and minimum-entropy optimization to calibrate interest-rate models.

```
 1  Needs[ "RB'MtgMath'" ];
 2
 3  c1 = ECALL[ 60, 95 ];
 4  c2 = ECALL[ 60, 100 ];
 5  c3 = ECALL[ 60, 110 ];
 6  c4 = ECALL[ 60, 115 ];
 7
 8  pf = PORTFOLIO[LONG[c1], SHORT[c2], SHORT[c3], LONG[c4]];
 9
10  v = VOL[ IMPLIED[ 60, {10 (* min *), 30 (* max *) } ] ];
11  r = DRIFT[ IMPLIED[ 60, 3 ] ];
12  s = SCENARIO[ WORSTCASE[ SELLER ] ];
13
14  sc = Script[ pf, v, r, s, AssetPrice -> 100,
15          TimeStep -> 0.1, Method -> explicit ];
16
17  Print[ RunScript[ sc ] ];
18
19     (* print output: *)
20  {  (* total value, delta, gamma, running time *)
21     {3.72652,0.350664,-0.0689839,0.44},
22     (* gradient w.r.t. c1, c2, c3, c4 *)
23     {5.61597,1.87094,0.019013,0.00049443},
24     (* today's time slice of the lattice, 3 points *)
25     {{99.7669,3.6429},{100,3.72652},{100.234,3.80657}} }
26
27  f = MakePriceFunc[ {S}, v, r, s, AssetPrice -> S,
28          TimeStep -> 0.1, Method -> explicit ];
29
30  (* f can be plotted as in:  Plot[f[S], {S,90,120}] *)
```

Fig. 1.2. A Mathematica notebook for MtgMath. The bufferfly spread with strikes 95, 100, 110 and 115 maturing in 60 days is worth 3.72652 when the asset price is 100 and the volatility range 10–30 %. The function f maps current asset price to worst-case portfolio value

1.5 How to Best Read This Book

The book progresses from a focus on theory and no implementation-related material to no theory and exclusively implementation-related chapters. Early chapters contain useful information on the theory of uncertain volatility. The middle part focuses on algorithms for barrier and American option portfolios, while the final chapters describe the actual implementation of those algorithms.

In more detail:

Chapter 2 summarizes notation and conventions, most of which are standard or intuitive. This chapter can be consulted as needed.

Chapter 3 gives a short overview over mathematical finance. The Black-Scholes model receives the most attention, although interest-rate models such as HJM are also mentioned. The distinction between deterministic and stochastic volatility is emphasized. This chapter can be skipped safely by anyone familiar with the terms.

Chapter 4 reviews the concept of uncertain model coefficients and introduces the notion of "scenarios." Pricing, hedging and calibration are briefly discussed as three applications. An understanding of these issues is essential.

Chapter 5 introduces the multi-lattice framework within which our algorithms are developed. This chapter defines key data structures that should not be missed. It also presents insight into some numerical issues regarding stability in Sect. 5.2. This section is rather technical and can be skipped at first reading (not by anyone actually implementing the algorithms of this book, though).

Chapter 7 discusses algorithms for scenario-based evaluation of barrier option portfolios. In particular, it shows how to set up multi-lattice dynamic programming so that the potentially large number of PDE's can be handled. This chapter and the next are the heart of the book. Please read them.

Chapter 8 discusses algorithms for scenario-based evaluation of American option portfolios. Evaluating American options is more complex than evaluating barrier options, but the same idea of ordering solutions of PDE's hierarchically applies. The economic implications are discussed in Sect. 8.1.

Chapter 9 describes an extension to worst-case volatility scenarios. The volatility is now allowed to exhibit short shocks at unpredictable times. This chapter is independent of Chapters 7 and 8 and can be read immediately after Chapter 5. The style is less formal.

Chapter 10 gives an overview over the C++ class library MtgLib, which forms the core of the software system Mtg. MtgLib is an implementation of the algorithms presented in Chapters 7, 8 and 9.

Chapter 11 describes those C++ classes in MtgLib that correspond to tangible concepts like "instrument", "model", "scenario", and which we thus call "external."

Chapter 12 describes C++ those classes in MtgLib that drive evaluations behind the scene, and which we thus call "internal."

Chapter 13 describes C++ extensions that make it possible to use MtgLib for the calibration of fixed-income models with Monte-Carlo. For historical reasons, MtgLib leans heavily towards lattice methods.

The appendix describes very technical aspects and is a useful source of information for readers who want to study the software on the accompanying CD.

Appendix A describes some technical aspects of MtgClt and MtgSvr, two components of a network pricer for option portfolios.

Appendix B defines the scripting language in which MtgClt and MtgSvr communicate.

Appendix C introduces a Mathematica API to MtgLib.

For a quick demonstration of the software on a Windows computer, click on or start the file

\Mtg\Win32\go.bat

on the CD.

For more information and updates, please visit

http://robertbuff.com/uvm

Part I

Computational Finance: Theory

2 Notation and Basic Definitions

2.1 Linear Algebra

$\mathbb{N}$ denotes the nonnegative integers. $\mathbb{R}$ denotes the real numbers. $\mathbb{R}_+$ denotes the nonnegative real numbers. $\mathbb{R}_{++}$ denotes the strictly positive real numbers.

Vectors and matrices are typeset in boldface (except when greek symbols are used): $\mathbf{x} \in \mathbb{R}^n$, $\mathbf{A} \in \mathbb{R}^{n \times m}$. Vectors are interpreted as column vectors: $\mathbb{R}^n = \mathbb{R}^{n \times 1}$. In text, they are quoted in transposed form.

Normal font is used for vector or matrix components: $\mathbf{a} = (a_1, \ldots, a_n)^\mathrm{T} = (a_i \mid 0 \leq i \leq n)^\mathrm{T}$.

The zero vector is denoted by $\mathbf{0}$. The dot product is written $\mathbf{a} \cdot \mathbf{b} = \mathbf{a}^\mathrm{T} \mathbf{b}$. If $\mathbf{a} \in \mathbb{R}^n$ is a vector, $\mathbf{M} = \mathbf{I_a}$ denotes the diagonal matrix $\mathbf{M} \in \mathbb{R}^{n \times n}$ with $m_{ii} = a_i$ and all off-diagonal elements zero. For $\mathbf{M} \in \mathbb{R}^{n \times n}$, the trace of $\mathbf{M}$ is the sum of its diagonal elements: $\mathrm{tr}\,(\mathbf{M}) = \sum_{i=1}^n m_{ii}$.

For $\mathbf{x}, \mathbf{y} \in \mathbb{R}^n$, $\mathbf{x} \geq \mathbf{y}$ means $x_i \geq y_i$ for $1 \leq i \leq n$. $\mathbf{x} > \mathbf{y}$ means $\mathbf{x} \geq \mathbf{y}$ and there is at least one $j \in \{1, \ldots, n\}$ such that $x_j > y_j$. $\mathbf{x} \gg \mathbf{y}$ means $x_i > y_i$ for $1 \leq i \leq n$ throughout.

2.2 Probability and Stochastic Processes

Let $(\Omega, \mathcal{F}, P)$ be a probability space. A family of σ-algebras $\{\mathcal{F}_t \mid t \geq t_0\}$ is called a *filtration* on $(\Omega, \mathcal{F})$ if $\mathcal{F}_t \subseteq \mathcal{F}_{t'} \subseteq \mathcal{F}$ for $t_0 \leq t \leq t'$. Here, both $t \in \mathbb{N}$ or $t \in \mathbb{R}_+$ are admissible. $(\Omega, \mathcal{F}, \{\mathcal{F}_t\}, P)$ is called a *filtered probability space*. We say it satisfies the *usual conditions* if $\mathcal{F}$ is P-complete, $\mathcal{F}_0$ contains all P-nullsets of $\mathcal{F}$ and $\{F_t\}$ is right-continuous.

Let $X = \{X_t \mid t_0 \leq t \leq t_1\}$ be a stochastic process defined on Ω. If the range of the index t is clear we write $\{X_t\}$. If the sample points X_t are random variables in $\mathbb{R}$ we write $X_t \in \mathbb{R}$. If the sample points $\mathbf{X}_t$ are n-vectors of random variables we write $\mathbf{X}_t \in \mathbb{R}^n$. Given $\omega \in \Omega$, we call $\{X_t(\omega) \mid t_0 \leq t \leq t_1\}$ the *sample path* of the process X on ω.

X is called *adapted* to the filtration $\{\mathcal{F}_t\}$ if the random variable X_t is $\mathcal{F}_t$-measurable for every t. The filtration $\{\mathcal{F}_t^X\} = \{\sigma\{X_s \mid s \leq t\} \mid t \geq t_0\}$ of the σ-algebra of X is called the *natural filtration* of X. If $\{\mathcal{F}_t\} = \{\mathcal{F}_t^X\}$ we say that $\{\mathcal{F}_t\}$ is *generated* by X. If a filtered probability space $(\Omega, \mathcal{F}, \{\mathcal{F}_t\}, P)$ appears without further comments, we assume that $(\Omega, \mathcal{F}, \{\mathcal{F}_t\}, P)$ satisfies the usual conditions, and the stochastic process X under consideration generates $\{\mathcal{F}_t\}$. In particular, we assume $\mathcal{F}_0 = \{\Omega, \emptyset\}$. These definitions apply to the discrete case $t, t_0, t_1 \in \mathbb{N}$ as well as to the continuous case $t, t_0, t_1 \in \mathbb{R}_+$. In the discrete case, i, j, k are the preferred index symbols (instead of t, u etc.).

For any event $A \in \mathcal{F}$, we write $P(A)$ for the probability of A under the measure P. For any random variable X, we write $\mathrm{E}_P(X)$ for the expectation of X under the measure P. If it is clear which measure is meant, we simply write $\mathrm{E}(X)$. Two measures P and Q are *equivalent* if they have the same nullsets. The indicator random variable for $A \in \mathcal{F}$ is denoted by 1_A.

These definitions can be found in [14, 56, 64] or any other textbook on stochastic calculus.

2.3 Partial Portfolios and Positions

Uncertain volatility models require us to look at portfolios rather than individual instruments. A portfolio consists of a set of instruments, characterized by their payoff, and positions in these instruments. Payoff functions are random variables, and positions are real numbers.

Let $\mathbf{X}$ be a vector of k random variables (i.e., a portfolio of k contingent claims), and let $\lambda \in \mathbb{R}^k$ be a position in $\mathbf{X}$. Let $M \subseteq \{1, \ldots, k\}$, and let $i_1 < i_2 < \cdots < i_n$ be an enumeration of the n elements in M. The selection operator on $\mathbf{X}$ and λ is defined as follows:

$$\begin{aligned}
\text{select}(\mathbf{X}, M) &= (X_{i_1}, \ldots, X_{i_n})^\mathrm{T} \\
\text{select}(\lambda, M) &= (\lambda_{i_1}, \ldots, \lambda_{i_n})^\mathrm{T}
\end{aligned} \quad (2.1)$$

A vector $\mathbf{Y}$ of $k' \leq k$ random variables is called a *restriction* or *partial portfolio* of $\mathbf{X}$, in symbolic notation $\mathbf{Y} \subseteq \mathbf{X}$, if there is $M \subseteq \{1, \ldots, k\}$, $|M| = k'$, such that $\mathbf{Y} = \text{select}(\mathbf{X}, M)$. $\mathbf{Y}$ results by removing some claims from $\mathbf{X}$ (possibly none!). The interpretation of "=" and "<" is straightforward.

The definition of a *restriction* or *partial position* $\lambda' \in \mathbb{R}^{k'}$ of λ is analogous: we write $\lambda' \subseteq \lambda$ if there is $M \subseteq \{1, \ldots, k\}$, $|M| = k'$, such that $\lambda' = \text{select}(\lambda, M)$.

We write $(\mathbf{Y}, \lambda') \subseteq (\mathbf{X}, \lambda)$ if $\mathbf{Y} \subseteq \mathbf{X}$ and $\lambda' \subseteq \lambda$ for the same $M \subseteq \{1, \ldots, k\}$, and call $(\mathbf{Y}, \lambda')$ a *partial portfolio* of $(\mathbf{X}, \lambda)$ (notice the re-use of the term).

For example, let $(\mathbf{X}, \lambda) = ((A, B), (1, -1))$ be a portfolio with a long position with payoff A and a short position with payoff B. Then $((A, B), (1, -1))$, $((A), (1))$, $((B), (-1))$ and $((), ())$ are all partial portfolios of $(\mathbf{X}, \lambda)$.

2.4 Accents, Superscript, Subscript

Let x be a variable that stands for a quantity, function, or stochastic process. We denote the value or instantiation of x that solves some optimization problem with $\hat{x}$. External values of x such as market prices of instruments are denoted $\bar{x}$.

Let $\mathbf{X}$ be a vector of k stochastic processes. Because time is regarded as the dominant index, subscripts denote the value of $\mathbf{X}$ at a given time t (or j, if time is discrete): $\mathbf{X}_t \in \mathbb{R}^k$, $\mathbf{X}_j \in \mathbb{R}^k$. Superscripts denote individual vector components: $\mathbf{X}_t = \{X_t^1, \ldots, X_t^k\}$. This differs from the notation for plain vectors where time plays no role (see above).

Let $S = \{S_t\}$ be a stochastic process. Its stochastic differential equation is written $dS = \mu S_t + \sigma\, dW$ (Black-Scholes being one example). In general, we omit the time subscript on the left side where the differential is taken, and include it on the right side where S_t is not differentiated.

3 Continuous Time Finance

This chapter gives a brief survey of continuous time finance. We categorize diffusion models according to the nature of their volatility coefficient. Models whose volatility coefficient does not exhibit randomness are treated in Sect. 3.1. Models whose volatility coefficient follows a stochastic process are discussed in Sect. 3.2.

The material presented in this chapter is standard. Uncertainty volatility models, on which the original work of this book is grounded, are discussed in Chapter 4.

3.1 Deterministic Volatility

Most of our work is based on equity/FX Black-Scholes models. For this reason, Black-Scholes analysis is reviewed in greater detail.

Since the software on the accompanying CD can also be used to calibrate interest rate models, we also describe some interest rate models.

3.1.1 One-Factor Black-Scholes Analysis

There are several ways to derive the Black-Scholes partial differential equation. References [10], [26], [43], [78], for example, all use stochastic calculus, and in particular Ito's formula (a good source for which is [64]). In [25] an alternative derivation is used: the Black-Scholes formula can also be interpreted as the continuous-time limit of a binomial random walk model.

Given is a filtered probability space $(\Omega, \mathcal{F}, \{\mathcal{F}_t\}, P)$ and a finite time horizon T. In this probability space, let $B = \{B_t\}$, $B_0 = 1$ be the price process of a riskless asset (B for bond), and let $S = \{S_t\}$ be the security price process:

$$\begin{aligned} dB &= r_t B_t \\ dS &= S_t(\mu_t \, dt + \sigma_t \, dW) \end{aligned} \tag{3.1}$$

W is a Brownian motion and r_t, μ_t and σ_t are sufficiently well-behaved functions. Let X be a nonnegative $\mathcal{F}_T$-measurable random variable that represents the payoff structure of a contingent claim on S.

3 Continuous Time Finance

$$\phi_t = \frac{\mu_t - r_t}{\sigma_t} \tag{3.2}$$

and

$$\zeta_t = \exp\left\{-\int_0^t \phi_u \, dW_u - \frac{1}{2}\int_0^t \phi_u^2 \, du\right\} \tag{3.3}$$

define a martingale measure Q equivalent to P via $Q(A) = \mathrm{E}_P(\zeta_t 1_A)$ for all $A \in \mathcal{F}$. The arbitrage-free price π of the contingent claim X is given by

$$\pi = \mathrm{E}_Q(\beta_T X) \tag{3.4}$$

where $\beta = \{\beta_t\}$, $\beta_t = 1/B_t$, is the discount process belonging to B.

In order to *compute* π, a replicating strategy for X is constructed explicitly. Let $f(S_t, t)$ denote the (yet unknown) price of X at time t for security price S_t^1, with final value $f(S_T, T) = X$. Let $F = \{F_t\}$ be the associated price process: $F_t = f(S_t, t)$. Assume for the moment that f is twice differentiable.

Define the $\mathbb{R}^2$-valued process $\theta = \{(\theta_t^0, \theta_t^1)\}$ of a position θ_t^0 in the riskless asset and a position θ_t^1 in the security as follows:

$$\theta_t^0 = \beta_t\left(F_t - \frac{\partial}{\partial S}f(S_t,t) S_t\right) \quad \text{and} \quad \theta_t^1 = \frac{\partial}{\partial S}f(S_t,t) \tag{3.5}$$

The portfolio implied by θ replicates F at all times and thus X at time T:

$$\theta_t^0 B_t + \theta_t^1 S_t = F_t \tag{3.6}$$

Now notice that, with Ito's formula,

$$dF = \left(\frac{\partial f}{\partial t} + \mu_t S_t \frac{\partial f}{\partial S} + \frac{1}{2}\sigma_t^2 S_t^2 \frac{\partial^2 f}{\partial S^2}\right) dt + \sigma_t S_t \frac{\partial f}{\partial S} dW \tag{3.7}$$

This implies together with the definition of θ that the instantaneous change of the value of the portfolio $(\theta_t^0 B_t, \theta_t^1 S_t)$ is

$$\begin{aligned}
&\theta_t^0 \, dB + \theta_t^1 \, dS = \\
&(\theta_t^0 r_t B_t + \theta_t^1 \mu_t S_t) \, dt + \theta_t^1 \sigma_t S_t \, dW = \\
&dF - \left(\frac{\partial f}{\partial t} + \frac{1}{2}\sigma_t^2 S_t^2 \frac{\partial^2 f}{\partial S^2} + r_t S_t \frac{\partial f}{\partial S} - r_t F_t\right) dt = \\
&dF - \rho \, dt
\end{aligned} \tag{3.8}$$

where ρ stands for the term in the brackets. θ is self-financing only if $\rho = 0$, in which case F is the value process corresponding to θ, and

$$\theta_t^0 \, dB + \theta_t^1 \, dS = dF \tag{3.9}$$

and

$$F_t - F_0 = \int_0^t \theta_u^0 dB + \theta_u^1 dS \, du \qquad (3.10)$$

hold.

It is the condition $\rho = 0$ which gives rise to the Black-Scholes partial differential equation

$$\frac{\partial f}{\partial t} + \frac{1}{2}\sigma_t^2 S_t^2 \frac{\partial^2 f}{\partial S^2} + r_t S_t \frac{\partial f}{\partial S} - r_t f_t = 0 \qquad (3.11)$$

with boundary condition

$$f(S_T, T) = X \qquad (3.12)$$

Fact 1. *If there is no arbitrage, then the price function $f: (0, \infty) \times [0, T] \to \mathbb{R}_+$ for X satisfies (3.11). In this case, (3.5) defines the replicating trading strategy.*

We have $f(S_0, 0) = \pi$ and therefore $f(S_0, 0) = E_Q(\beta_T X)$. Moreover,

$$F_t = \frac{1}{\beta_t} E_Q(\beta_T X \mid \mathcal{F}_t) \qquad (3.13)$$

This is sometimes called the *probabilistic solution* of (3.11).

3.1.2 Hedging with Black-Scholes

There is an intuitive economic interpretation of (3.11): the difference of the return of a hedged option portfolio (the first two terms) and a bank account (the last two terms) must be zero. The derivation of (3.11) shows the dominant role of σ_t; the drift μ_t, on the other hand, can be "hedged away".

Delta-Hedging The replicating portfolio (3.5) consists of a position in cash and a position in the underlying asset. The self-financing replicating strategy requires to dynamically adjust those positions. Rearranging the terms of (3.9), using the definition of θ_t^0 in (3.5), dB in (3.1), and the fact that $\beta_t B_t = 1$ we get

$$\begin{aligned} dF - \theta_t^1 \, dS = dF - \frac{\partial f}{\partial S} dS &= \theta_t^0 \, dB \\ &= (F_t - \frac{\partial f}{\partial S} S_t) r_t \end{aligned} \qquad (3.14)$$

The last term represents the appreciation of a portfolio that consists of the option and a hedging position in the asset. Equation (3.14) demonstrates the hedging aspect of Black-Scholes analysis.

Fact 2. Let $f: (0, \infty) \times [0, T] \to \mathbb{R}_+$ be the option price function. A position in the option can be hedged by dynamically holding $-\frac{\partial}{\partial S} f(S_t, t)$ units of the underlying asset. The value of the combined portfolio of the asset and the option appreciates at the risk-free interest rate.

This strategy is called delta hedging. The "delta" of the option is

$$\Delta = \frac{\partial f}{\partial S} \tag{3.15}$$

The delta of the hedged position is zero; the position is "delta-neutral." The cost of delta hedging is covered entirely by the option premium $f(S_0, 0)$.

Higher Order Derivatives The second-order derivative of an option or a portfolio with respect to moves of the asset price is called Γ ("gamma"):

$$\Gamma = \frac{\partial^2 f}{\partial S^2} = \frac{\partial \Delta}{\partial S} \tag{3.16}$$

Γ plays an important part in uncertain volatility models.

Theta, Kappa, Rho Theta denotes the time decay of the option or portfolio value:

$$\Theta = -\frac{\partial f}{\partial t} \tag{3.17}$$

Kappa denotes the volatility risk:

$$\kappa = \frac{\partial f}{\partial \sigma} \tag{3.18}$$

Rho denotes the interest rate risk:

$$\varrho = \frac{\partial f}{\partial r} \tag{3.19}$$

From (3.14) and (3.11) we find that for a delta-hedged portfolio,

$$\begin{aligned} dF - \Delta\, dS &= (F_t - \Delta\, dS)\, r_t \\ &= \left(\frac{\partial f}{\partial t} + \frac{1}{2} \sigma_t^2 S_t^2 \frac{\partial^2 f}{\partial S^2} \right) dt \\ &= \left(-\Theta + \frac{1}{2} \sigma_t^2 S_t^2 \Gamma \right) dt \end{aligned} \tag{3.20}$$

The larger Γ, the stronger time decay, and vice versa.

See [25] for detailed material on the sensitivities of put and call option prices to Black-Scholes parameters.

3.1 Deterministic Volatility

Hedging with Other Assets In practice r and σ are obtained from prices of other liquid assets. For example, it may make sense to choose r such that the N discounted coupon and principal cashflows $C_1, \ldots, C_N$ with maturities $T_1, \ldots, T_N$ of a given reference bond reproduce its market price p which we know. We want the following condition to hold:

$$p = P(r) = \sum_{i=1}^{N} e^{-rT_i} C_i \qquad (3.21)$$

In order to hedge rho for an option contract we must trade the reference bond; the interest rate r is not directly tradable.

Let $f = f(S_t, t; r)$ be the price function of the option, with the dependency on r emphasized. In order to be rho-neutral we must find θ_p such that

$$\frac{\partial f}{\partial r} + \theta_p \frac{\partial P}{\partial r} = 0 \qquad (3.22)$$

or

$$\theta_p = -\frac{\partial f}{\partial r} \bigg/ \frac{\partial P}{\partial r} \qquad (3.23)$$

θ_r is the number of bond units we must hold.

The hedge ratio θ_p can also be computed directly:

$$\begin{aligned}
\theta_p &= -\frac{\partial f}{\partial p} = -\frac{\partial f}{\partial r}\frac{\partial r}{\partial p} = -\frac{\partial f}{\partial r}\frac{\partial}{\partial p} P^{-1}(p) \\
&= -\frac{\partial f}{\partial r} \bigg/ \frac{\partial}{\partial r} P(P^{-1}(p)) = -\frac{\partial f}{\partial r} \bigg/ \frac{\partial P}{\partial r}
\end{aligned} \qquad (3.24)$$

Elementary calculus thus leads to the same result.

3.1.3 Interest Rate Models

Interest-rate derivatives can in some sense be regarded as a bet on the future cost of money. Instead of a stock price process S we find stochastic processes of bond prices, yields, short or forward rates, depending on the focus of the model.

Terminology Let $(\Omega, \mathcal{F}, \{\mathcal{F}_t\}, P)$ be the underlying filtered probability space. $\mathbf{W} = \{\mathbf{W}_t\}$ is an N-dimensional Brownian motion on it; τ a finite time horizon. In this context, the symbol T usually denotes the maturity of a bond in the literature. We follow this convention here, but use T for other purposes in later sections.)

Assume a continuum of *discount bonds*, one for each maturity $T \leq \tau$. The time t-price of the bond with maturity T is denoted by $P(t, T)$, with terminal price $P(T, T) = 1$ (all bonds are normalized). The *instantaneous forward rate*

20   3 Continuous Time Finance

at time t for borrowing at time T, $f(t,T)$, and the *yield*—the average implied interest rate—at time t of the bond maturing at time T, $R(t,T)$, fulfill

$$f(t,T) = -\frac{\partial}{\partial T} \log P(t,T)$$
$$R(t,T) = -\frac{\log P(t,T)}{T-t}$$
(3.25)

for all $0 \leq t < T \leq \tau$, respectively. Solving for P, one gets

$$P(t,T) = \exp\left(-\int_t^T f(t,u)\, du\right) \tag{3.26}$$

The time t instantaneous forward rate, defined as

$$r_t = f(t,t) \tag{3.27}$$

is called the *short rate*. Note that the short rate is not sufficient to recover $P(t,T)$; the entire forward rate curve is needed.

It is assumed that there exists a *cash bond* $B = \{B_t\}$, whose stochastic differential equation is

$$dB = r_t B_t\, dt \tag{3.28}$$

B is the numeraire. With $B_0 = 1$, the solution of (3.28) is

$$B_t = \exp\left(\int_0^t r_u\, du\right) \tag{3.29}$$

for $0 \leq t \leq \tau$. Again, we define a discount factor $\beta_t = 1/B_t$.

The Heath-Jarrow-Morton Framework (HJM) The HJM framework first presented in [37] defines generic behavior that every no-arbitrage interest rate model should observe. In its most general formulation the entire forward rate curve is evolved, starting with a term-structure of interest rates observed in today's market.

An approach based on discrete trees is used in [51].

For $0 \leq T \leq \tau$, let the $\mathbb{R}$-valued process $f^T = \{f_t^T\}$ denote the evolution of the time t forward rate for borrowing at time T: $f_t^T = f(t,T)$. The dynamics of the f^T are

$$f_t^T = f_0^T + \int_0^t \alpha_u^T(\omega)\, du + \sum_{i=1}^N \int_0^t \sigma_u^{Ti}(\omega)\, dW^i \tag{3.30}$$

for $0 \leq t \leq T$. Here $\{f_0^T = f(0,T) \mid 0 \leq T \leq \tau\}$ is a non-random initial forward rate curve, and the $\mathbb{R}$-valued processes α^T and σ^{Ti} ($0 \leq T \leq \tau$, $1 \leq i \leq N$) may depend on ω, are adapted to $\{\mathcal{F}_t\}$ and satisfy certain continuity, integrability and boundedness conditions. We will omit the argument ω to enhance readability. (In the literature, f_t^T is usually written $f(t,T)$, α_t^T is written $\alpha(t,T)$ and σ_t^{Ti} is written $\sigma_i(t,T)$. In order to be consistent with our earlier notation, we keep the current time t as a subscript, the index of the asset as first superscript and the index of the source of uncertainty as the second superscript.)

With (3.27) and (3.30), the short rate process can be written as

$$r_t = f_0^t + \int_0^t \alpha_u^t \, du + \sum_{i=1}^N \int_0^t \sigma_u^{ti} \, dW^i \tag{3.31}$$

Ito's lemma together with some regularity conditions on B show that $P(t,T)$ is the solution of

$$dP(t,T) = P(t,T)\left[(r_t + b_t^T)dt + \sum_{i=1}^N a_t^{Ti} dW^i\right] \tag{3.32}$$

with

$$\begin{aligned}
a_t^{Ti} &= -\int_t^T \sigma_u^{Ti} \, du \quad (1 \le i \le N) \\
b_t^T &= -\int_t^T \alpha_u^T \, du + \frac{1}{2}\sum_{i=1}^N (a_t^{Ti})^2
\end{aligned} \tag{3.33}$$

b_t^T is the excess rate of return of the T-maturity bond at time t. The bond price processes $P(t,T)$ are not necessarily Markovian!

Under no-arbitrage assumptions, it is necessary to find an equivalent measure Q which makes the discounted bond price processes $\beta_t P(t,T)$ martingales, simultaneously for all $0 \le T \le \tau$. [37] argue that it is sufficient to find such Q for a "basis" of N different bonds. It can furthermore be shown that Q, if it exists, is unique and does not depend on the choice of the basis.

After doing this, the short rate r_t follows the process

$$r_t = f_0^t + \sum_{i=1}^N \int_0^t \sigma_u^{ti} \int_u^t \sigma_v^{vi} \, dv \, du + \sum_{i=1}^N \int_0^t \sigma_u^{ti} \, d\tilde{W}^i \tag{3.34}$$

where $\tilde{W}$ is a Q-Brownian motion. In general, r_t is path dependent. Note that the drift α does not appear in (3.34).

Given the martingale measure Q, contingent claims X that mature at some time T are evaluated in standard fashion, with fair price $\pi = E_Q(\beta_T X)$ and value process

$$X_t = \frac{1}{\beta_t} E_Q(\beta_T X \mid \mathcal{F}_t) = E_Q\left(\exp\left(-\int_t^T r_u \, du\right) X \mid \mathcal{F}_t\right) \tag{3.35}$$

In particular,

$$P(t,T) = E_Q\left(\exp\left(-\int_t^T r_u\,du\right)\bigg|\mathcal{F}_t\right) \tag{3.36}$$

The HJM model is a general framework in which many other interest rate models can be re-expressed by instantiating drift and volatility coefficients in specific ways. We list some other well-known interest rate models in the following paragraphs. The reader who wants to study interest rate models further is referred to [50].

We start our survey with Gaussian and CIR models. The models can can have negative interest rates with positive probability.

The Vasicek Short-Rate Model In, [80] a one-factor model is proposed where the short rate follows an Ornstein-Uhlenbeck mean reverting process under the equivalent martingale measure Q:

$$dr = (\theta - \alpha r_t)dt + \sigma\,d\tilde{W} \tag{3.37}$$

with constants θ, α and σ. In terms of HJM, this means

$$\sigma_t^T = \sigma e^{-\alpha(T-t)}$$
$$f_0^T = \theta/\alpha + e^{-\alpha T}(r_0 - \theta/\alpha) - \frac{\sigma^2}{2\alpha^2}(1 - e^{-\alpha T})^2 \tag{3.38}$$

The Extended Vasicek Short-Rate Model The extended Vasicek model is obtained by making the drift, mean reversion and volatility coefficients of the Vasicek model deterministic time-dependent functions:

$$dr = (\theta_t - \alpha_t r_t)dt + \sigma_t\,d\tilde{W} \tag{3.39}$$

It can be shown (for instance, see [50]) that

$$R(t,T) = \frac{A(t,T)}{T-t} + \frac{B(t,T)}{T-t}r_t \tag{3.40}$$

with

$$A(t,T) = \int_t^T \frac{1}{K(u)}\int_t^u \left(\theta_s K(s) - \frac{1}{K(s)}\int_t^s \sigma_y^2 K^2(y)\,dy\right)ds\,du$$
$$B(t,T) = K(t)\int_t^T \frac{1}{K(u)}\,du \tag{3.41}$$
$$K(t) = \exp\left(\int_0^t \alpha_s\,ds\right)$$

With (3.25) and (3.40) one finds that zero-coupon bond prices $P(t,T)$ in the extended Vasicek model satisfy

$$P(t,T) = \exp\left(-A(t,T) - B(t,T)\, r_t\right) \tag{3.42}$$

If $A(t,T)$ and $B(t,T)$ are stationary, i.e. $A(t,T) = A(t+dt, T+dt)$ and $B(t,T) = B(t+dt, T+dt)$ then the model in (3.39) is affine. For example, the Vasicek model with constant coefficients is affine.

The Hull and White Model In [47] the mean-reversion and volatility coefficients of the extended Vasicek model are defined constant:

$$\mathrm{d}r = (\theta_t - \alpha\, r_t)\, \mathrm{d}t + \sigma\, \mathrm{d}\tilde{W} \tag{3.43}$$

The Ho and Lee Model This model first presented in [42] has no mean reversion:

$$\mathrm{d}r = \theta_t\, r_t\, \mathrm{d}t + \sigma\, \mathrm{d}\tilde{W} \tag{3.44}$$

The Extended Cox-Ingersoll-Ross Model The original model was proposed in [23] and [24] and has constant coefficients. Members of this model family have the general form

$$\mathrm{d}r = (\theta_t - \alpha_t\, r_t)\mathrm{d}t + \sigma_t\, \sqrt{r_t}\, \mathrm{d}\tilde{W} \tag{3.45}$$

For the general version,

$$R(t,T) = \frac{A(t, T-t)}{T-t} + \frac{B(t, T-t)}{T-t}\, r_t \tag{3.46}$$

where $A(t, T-t)$ and $B(t, T-t)$ satisfy ordinary differential equations:

$$\begin{aligned}\frac{\partial A(t, T-t)}{\partial t} &= -\theta_t B(t, T-t) \\ \frac{\partial B(t, T-t)}{\partial t} &= 1 + \alpha_t B(t, T-t) - \frac{1}{2}\sigma_t^2 B(t, T-t)^2\end{aligned} \tag{3.47}$$

See [50] or [43] for closed-form solutions for the constant case.

Models that guarantee positive interest rates are called positive interest rate models. We list two log-normal models which fall into this category.

The Black-Karasinski Model This model, presented in [13], has the form

$$\mathrm{d}X = (\theta_t + \alpha_t\, X_t)\mathrm{d}t + \sigma_t\, \mathrm{d}\tilde{W} \tag{3.48}$$

$X = \{X_t\}$ is the Gaussian stochastic process of an abstract, non-observable parameter. The short rate is defined as

$$r_t = \mathrm{e}_t^X \tag{3.49}$$

The Black, Derman and Toy Model This earlier model, presented in [12], has the form

$$dX = \left(\theta_t + \frac{1}{\sigma_t}\frac{\partial \sigma}{\partial t} X_t\right) dt + \sigma_t d\tilde{W} \qquad (3.50)$$

The short rate is defined as

$$r_t = e^{X_t} \qquad (3.51)$$

Implementors must pay special attention: when defined as above, the following process becomes unbounded:

$$V_t = \mathrm{E}\left(\exp\left(\int_0^t r_s \, ds\right)\right) \qquad (3.52)$$

3.2 Stochastic Volatility

In Section 3.1 the volatility coefficient was either constant or a deterministic function of time and thus independent of the current level of the underlying stochastic process. A generalized volatility coefficient of the form $\sigma(S_t, t)$ as in

$$dS = \mu(S_t, t)\, dt + \sigma(S_t, t)\, dW \qquad (3.53)$$

is called *level-dependent*. Because volatility and asset price are perfectly correlated, we still have only one source of randomness: $W = \{W_t\}$. A time and/or level-dependent volatility coefficient makes the arithmetic more challenging and commonly precludes the existence of a closed-form solution. However, the arbitrage argument based on portfolio replication and a complete market still goes through unchanged. The situation is different if the volatility is influenced by a second "nontradable" source of randomness and we have *stochastic volatility*.

The folloing pages give an overview over stochastic volatility as an alternative to uncertain volatility, which we study in depth in Chapter 4.

3.2.1 Tradable and Nontradable Factors

The following overview follows [38]. Their work draws from results in [28]. The model is general enough to include the deterministic models of Sect. 3.2.2 as special cases. For a survey on level and stochastic volatility models the reader is also referred to [31].

Let $\mathbf{W} = \{\mathbf{W}_t\}$ be an N-dimensional Brownian motion on a filtered probability space $(\Omega, \mathcal{F}, \{\mathcal{F}_t\}, P)$. Fix some time horizon T. Define the $\mathbb{R}^M$-valued process $\mathbf{X} = \{\mathbf{X}_t\}$ with component processes $X^1, \ldots, X^M$ by

$$d\mathbf{X} = \mu(\mathbf{X}_t, t)\,dt + \sigma(\mathbf{X}_t, t)\,d\mathbf{W} \tag{3.54}$$

where $\mu = (\mu^i : \mathbb{R}^M \to \times [0,T]\mathbb{R} \mid 1 \leq i \leq M)$ and $\sigma = (\sigma^{ij} : \mathbb{R}^M \times [0,T] \to \mathbb{R} \mid 1 \leq i \leq M, 1 \leq j \leq N)$ are functions satisfying appropriate regularity conditions. The component processes $X^1, \ldots, X^M$ may represent tradable assets or economic factors; trivially, there must be at least one tradable asset X^i and we assume $i = 1$ without loss of generality.

We also postulate the existence of an $\mathbb{R}_{++}$-valued process X^0 which plays the role of the riskless asset:

$$dX^0 = r(\mathbf{X}_t, t) X^0_t\, dt \quad \text{and} \quad X^0_0 = 1 \tag{3.55}$$

The discount factor $\beta = \{\beta_t\}$ is defined via $\beta_t = 1/X^0_t$, as usual.

Let the random variable Y on $(\Omega, \mathcal{F}_T)$ be a contingent claim. Standard procedure would imply a replicating strategy $\theta = \{\theta_t\}$, $\theta_t \in \mathbb{R}^{M+1}$, for Y and a value process V^θ, which would then satisfy

$$V^\theta_t = \frac{1}{\beta_t} E_Q\left(\beta_T Y \mid \mathcal{F}_t\right) \tag{3.56}$$

under some P-equivalent measure $Q \in \mathbb{P}$ which makes $\beta \mathbf{X}$ a martingale.

This is indeed the case if the economy is complete, i.e. $N = M$ and all components are tradable. In this case, θ exists and is self-financing, and Q and θ are uniquely determined by μ, σ and Y_T.

In the general, incomplete situation, this need not be so. We would certainly wish $\theta^i \equiv 0$ to hold for all nontradable components i. However, this restriction might make a self-financing replicating strategy impossible. There are several ways out of this dilemma. In [71], for instance, "mean-self financing" strategies are discussed. Here, we briefly summarize some more concrete solutions.

3.2.2 Some Concrete One-Dimensional Models

We present some concrete models based on a ome-dimensional asset price process and stochastic volatility. The models differ in how they supplement no-arbitrage theory. Hull-White and Wiggins, for instance, advance equilibrium arguments, while Scott introduces a second derivative to exploit ad-hoc opportunities. See [9] for a discussion of the models by Hull and White, Stein and Stein, and Heston.

Let $(\Omega, \mathcal{F}, \{\mathcal{F}_t\}, P)$ be a filtered probability space and T a fixed, finite time horizon. Let $\mathbf{W} = \{\mathbf{W}_t\}$, $\mathbf{W}_t \in \mathbb{R}^2$, be a two-dimensional Brownian motion with correlation coefficient ρ, or $E_P\left(dW^1\, dW^2\right) = \rho\, dt$. (At this point, we deviate from our standard assumption that the component processes of $\mathbf{W}$ are independent, i.e. $\rho = 0$.)

There is a riskless asset $X^0 = \{X^0_t\}$ with $X^0_0 = 1$ and $X^0_t = e^{rt}$. r is the riskless rate and $\beta = 1/X^0$ the discount process, as usual.

26   3 Continuous Time Finance

Hull and White's Model [44] propose the following model:

$$\begin{aligned} dS &= S_t(r\,dt + \sigma_t\,dW^1) \\ d\sigma^2 &= \sigma_t^2\left(\xi(\sigma_t^2,t)\,dt + \phi(\sigma_t^2,t)\,dW^2\right) \end{aligned} \tag{3.57}$$

where ξ and ϕ may depend on σ^2 and t, but not on S. Under the additional assumption that (a) $\rho = 0$ and (b) σ^2 does not have systematic risk (a statement we shall not explain further at this point), a partial differential equation slightly more complex than (3.11) can be derived by using CAPM equilibrium arguments, eliminating randomness and therefore precluding any risk preferences.

Now define the mean volatility V over a particular path $\{\sigma_t^2\}$ as

$$V = \frac{1}{T}\int_0^T \sigma_t^2\,dt \tag{3.58}$$

For any attainable contingent claim X, let

$$\pi(V) = E_P\left(\beta_T X \mid \sigma^2 \equiv V\right) \tag{3.59}$$

denote the fair price of X under the restricted scenario where $\sigma_t^2 = V$ for $0 \le t \le T$. Then it can be shown that the no-arbitrage price of π is

$$\pi = \int_{-\infty}^{\infty} \pi(V)\,h(V \mid \sigma_0^2)\,dV \tag{3.60}$$

where $h(V \mid \sigma_0^2)$ is the density of V conditional on σ_0^2 under P. In other words, the price of a contingent claim X turns out to be the weighted average Black-Scholes price for any realizable mean volatility. This result does not hold if $\rho \ne 0$ or ξ or ϕ depend on S.

Wiggins' Model The model advocated in [77] has the dynamics

$$\begin{aligned} dS &= S_t(\mu\,dt + \sigma_t\,dW^1) \\ d\sigma &= h(\sigma_t)\,dt + \phi\sigma_t\,dW^2 \end{aligned} \tag{3.61}$$

It is not required that $\rho = 0$. Let $F = \{F_t\}$, $F_t = f(S_t,t)$, be the value process of a contingent claim X. Wiggins defines a hedge portfolio $\theta = \{\theta_t\}$, $\theta_t \in \mathbb{R}^2$, in the riskless asset and the security by

$$\begin{aligned} \theta_t^0 &= \beta_t\left(F_t - \frac{\partial}{\partial S}f(S_t,t)S_t - \rho\phi\frac{\partial}{\partial \sigma}f(S_t,t)\right) \\ \theta_t^1 &= \frac{\partial}{\partial S}f(S_t,t) + \frac{\rho\phi}{S_t}\frac{\partial}{\partial \sigma}f(S_t,t) \end{aligned} \tag{3.62}$$

θ is a modification of (3.5) with the property that its value process V^θ satisfies

$$\frac{dV^\theta}{V_t^\theta} \frac{dS}{S_t} = 0 \tag{3.63}$$

for $0 \leq t \leq T$. I.e., the return of the hedge portfolio is uncorrelated with the return of the security. If S is an index on the market, the hedge portfolio has therefore a zero beta coefficient. Under some additional economic assumptions and for the special case that S is indeed a contingent claim on the market, f is a solution to the partial differential equation

$$\text{BS} + \frac{1}{2}\sigma_t^2\phi^2 \frac{\partial^2 f}{\partial \sigma^2} + \rho\phi\sigma_t^2 S_t \frac{\partial f^2}{\partial S \partial \sigma} + \frac{\partial f}{\partial \sigma}\left(h(\sigma_t) - \rho\theta\sigma_t^2\right) = 0 \tag{3.64}$$

where BS stands for the left side terms of the Black-Scholes partial differential equation (3.11).

Johnson and Shanno's Model [53] choose the model

$$\begin{aligned} dS &= S_t(\mu\,dt + \sigma_t S_t^{\alpha_1 - 1}\,dW^1) \\ d\sigma &= \sigma_t(\xi\,dt + \phi\sigma_t^{\alpha_2 - 1}\,dW^2) \end{aligned} \tag{3.65}$$

with $\alpha_1, \alpha_2 \geq 0$. The correlation coefficient between W^1 and W^2 is ρ. Setting up the Black-Scholes replicating portfolio θ as in (3.5), one finds that the value process V^θ of θ satisfies

$$dV^\theta = dF_t - (\text{BS} + \text{JS})\,dt - \phi\sigma_t^{\alpha_2} \frac{\partial f}{\partial \sigma}\frac{\partial f}{\partial S}\,dW^2 \tag{3.66}$$

where BS represents the standard Black-Scholes terms—see (3.11)—and JS stands for additional nonrandom terms which are easy to derive with Ito calculus. At this point, Johnson and Shanno assume that the dW^2 term can be diversified away (this assumption replaces the equilibrium principles in the previous two models), and get a partial differential equation $\text{BS} + \text{JS} = 0$ with appropriate boundary conditions for X.

Scott's Model [72] uses a model in which the volatility follows a mean reverting process with mean $\bar{\sigma}$:

$$\begin{aligned} dS &= S_t(\mu\,dt + \sigma_t\,dW^1) \\ d\sigma &= \xi(\bar{\sigma} - \sigma)\,dt + \phi\,dW^2 \end{aligned} \tag{3.67}$$

Again, ρ is the correlation coefficient between dW^1 and dW^2. Assume there are two contingent claims, X and Y, with price functions f and g, respectively, and price processes $F = \{F_t\}$ and $G = \{G_t\}$. Assume furthermore that X expires at time $T_X \leq T$, and Y expires at time $T_X < T_Y \leq T$. A trading strategy θ that hedges a portfolio of X and Y (with dynamic weights) during times $0 \leq t \leq T_X$ gives rise to a partial differential equation

$$\frac{\partial g}{\partial \sigma}\left(\mathrm{BS}_f + \rho\phi\sigma_t S_t \frac{\partial f^2}{\partial S \partial \sigma} + \frac{1}{2}\phi^2 \frac{\partial f^2}{\partial \sigma^2}\right)$$
$$- \frac{\partial f}{\partial \sigma}\left(\mathrm{BS}_g + \rho\phi\sigma_t S_t \frac{\partial g^2}{\partial S \partial \sigma} + \frac{1}{2}\phi^2 \frac{\partial g^2}{\partial \sigma^2}\right) = 0 \qquad (3.68)$$

which does no longer have terms in dW^1 or dW^2. BS_f and BS_g represent the standard Black-Scholes terms corresponding to f and g terms as they appear in (3.11).

However, the PDE in (3.68) does not have a unique solution for given boundary conditions at $t = T_X$. There are two ways in which this situation can be resolved:

- Equilibrium arguments can be applied. This approach is chosen in [72] and leads to a partial differential equation for X which depends on λ^*, the risk premium associated with $d\sigma$.
- If the price for the claim Y is known (for instance, if Y is a liquid option), one can model Y's price process G as a geometric Brownian motion diffusion with the constant volatility implied by Y's price. This path is explored—based on a slightly different model for $d\sigma$—in [81].

A theoretical third possibility is to postulate the existence of an asset whose price is perfectly correlated with $d\sigma$.

The Model by Stein and Stein In [73], a mean reverting Ornstein-Uhlenbeck process for volatility is used as well:

$$dS = S_t(\mu\,dt + \sigma_t\,dW^1)$$
$$d\sigma = \xi(\bar{\sigma} - \sigma_t)\,dt + \phi\,dW^2 \qquad (3.69)$$

The level of mean reversion is $\bar{\sigma}$. In this model, dW^1 and dW^2 are independent, i.e. $\rho = 0$. Now let Φ be the market price of the volatility risk of S. Under certain assumptions similar to those used in [77] the pricing measure for options on S is implicitly given by

$$dS' = S'_t(r\,dt + \sigma'_t\,dW^1)$$
$$d\sigma' = \xi\left(\bar{\sigma}\frac{\Phi\phi}{\xi} - \sigma'_t\right)dt + \phi\,dW^2 \qquad (3.70)$$

Since the process for σ is arithmetic, not geometric as in (3.61), the authors are able to obtain a closed-form solution for the distribution of S at the desired time horizon.

Heston's Model In [41] the stock price return process and the variance process are modelled as

$$dS = S_t(\mu\,dt + \sqrt{v_t}\,dW^1)$$
$$dv = \xi(\bar{v} - v_t)\,dt + \phi\sqrt{v_t}\,dW^2 \qquad (3.71)$$

3.2 Stochastic Volatility   29

where $\sigma_t = \sqrt{v_t}$ is the volatility of the stock price return process. This formulation has the advantage of strictly positive volatility as long as $\xi \bar{v} \geq \phi^2/2$.

Let $F = \{F_t\}$, $F_t = f(S_t, t)$, be the value process of a contingent claim X on the stock. Heston finds that f satisfies the partial differential equation

$$\text{BS} + \frac{1}{2}\phi^2 v_t \frac{\partial^2 f}{\partial v^2} + \rho\phi v_t \frac{\partial f^2}{\partial S \partial v} + \frac{\partial f}{\partial v}\left(\xi(\bar{v} - v_t) - \lambda_t(S, v_t)\right) = 0 \quad (3.72)$$

Substitute the standard terms of the Black-Scholes partial differential equation (3.11) for "BS."

The term $\lambda_t(S, v_t)$ represents the market price of volatility risk. Heston argues that the choice

$$\lambda_t(S, v_t) = \lambda v_t \quad (3.73)$$

makes economic sense and derives a closed form solution of (3.72) for a European call option.

Fast Mean Reversion Model Let $f(S_t, t)$ be the Black-Scholes pricing function with constant volatility for a European option maturing at time T. It is shown in [30] that the price of the option adjusted for stochastic, fast mean reverting volatility can be expressed as

$$f(S_t, t) - (T - t)\left(V_2 S_t^2 \frac{\partial^2 f}{\partial S^2} + V_3 S_t^3 \frac{\partial^3 f}{\partial S^3}\right) \quad (3.74)$$

where the market constants V_2 and V_3 are obtained by fitting implied volatilities of liquid options.

Fast mean reversion emphasizes the burstiness of volatility: periods of high volatility tend to alternate with periods of low volatility. For example, consider the Ornstein-Uhlenbeck process

$$d\sigma = \xi(\bar{\sigma} - \sigma_t)\,dt + \phi\,dW \quad (3.75)$$

used in the model by Stein and Stein. The parameter ξ determines the speed of mean reversion: the larger its value the faster σ is driven back to the long-term mean $\bar{\sigma}$.

Recall the definition of the random variable V in (3.58) as the average volatility over any given volatility path. Because (3.75) is an ergodic process,

$$V = \frac{1}{T}\int_0^T \sigma_t^2\,dt \quad (3.76)$$
$$\to V_\infty \in \mathbb{R}$$

as $\xi \to \infty$. In other words: for a European option, extreme burstiness of volatility makes the stochastic volatility model and the non-stochastic Black-Scholes model evaluated at constant volatility $\sigma^2 = V_\infty$ indistinguishable.

The market parameters V_2 and V_3 introduced in (3.74) vanish for constant volatility, in which case the formula reverts to Black-Scholes. Non-zero V_2 and V_3 indicate less-than-infinite burstiness of volatility. The derivation of (3.74) in [30] assumes that volatility is always fast mean reverting.

3.3 Model Calibration

How should model coefficients be chosen? Risk-neutral pricing eliminates the drift coefficient, but volatility and mean reversion coefficients still remain. The model needs to be *calibrated*. This can be done in two ways:

– The model coefficients can be estimated from historical data.
– *Implied* coefficients are chosen such that the model replicates market prices of liquidly traded, often non-derivative assets. These assets are the model *benchmarks*.

In this book we focus on the implied method.

3.3.1 Parametric Methods

Let $S = \{S_t(\alpha)\}$ denote the stochastic price process of an underlying asset with model coefficient α. Let the random variables $\bar{X}_1, \ldots, \bar{X}_N$ denote the payoffs of N benchmark assets on the underlying, with maturities $T_1 < \cdots < T_N$ and market prices $\bar{\pi}_1, \ldots, \bar{\pi}_N$. Let $f_1(S_t, t; \alpha), \ldots, f_N(S_t, t; \alpha)$ be the benchmarks' model pricing functions (for example, obtained by solving the Black-Scholes partial differential equation (3.11) for each benchmark), given a particular choice of α.

Let X be the payoff of an additional derivative asset, with maturity T, $T_1 \leq T \leq T_N$, and let $f(S_t, t; \alpha)$ be its pricing function, given α. Parametric calibration and evaluation proceeds in these steps:

1. Find $\bar{\alpha}$ such that $f_i(S_0, 0; \bar{\alpha}) = \bar{\pi}_i$, $1 \leq i \leq N$.
2. Compute $\pi = f(S_0, 0; \bar{\alpha})$, the implied price of the derivative asset.

This method is called parametric. α usually has a simple form such as a constant or a deterministic function of time. In a computer program, information about the benchmark assets can be discarded after α has been computed.

Implied Volatility Assume $S = \{S_t\}$ is the Black-Scholes stock price process with constant coefficients:

$$dS = S_t(\mu \, dt + \sigma \, dW) \qquad (3.77)$$

The only coefficient in (3.77) affecting the prices of options on S is σ, i.e. we select $\alpha = \sigma$.

Let $N = 1$ and $\bar{X}_1$ be an option on S, maturing at time $T_1 \geq T$. Because $\bar{X}_1$ is liquidly traded, there is a market price $\bar{\pi}_1$. The volatility $\bar{\sigma}$ that satisfies

$$f_1(S_0, 0; \bar{\sigma}) = \bar{\pi}_1 \qquad (3.78)$$

is called *implied volatility*. The implied price of the illiquid option X is then

$$\pi = f(S_0, 0; \bar{\sigma}) \qquad (3.79)$$

Time-Dependent Coefficients Let $\bar{\pi}_1, \ldots, \bar{\pi}_N$ be the prices of N benchmark assets with maturities $T_1 < \cdots < T_N$. Let $f_1(S_t, t; \alpha), \ldots, f_N(S_t, t; \alpha)$ be the pricing functions with a time-dependent model coefficient $\alpha = \alpha(t)$.
Partition α into N functions as follows:

$$\alpha_1(t) = \begin{cases} \alpha(t) & \text{if } t \leq T_1 \\ 0 & \text{otherwise} \end{cases}$$

$$\alpha_i(t) = \begin{cases} \alpha(t) & \text{if } T_{i-1} < t \leq T_i \\ 0 & \text{otherwise} \end{cases} \quad (2 \leq i \leq N)$$

(3.80)

The implied $\bar{\alpha}$ can be constructed as follows:

1. Find $\bar{\alpha}_1$ such that

$$f_1(S_0, 0; \bar{\alpha}_1) = \bar{\pi}_1 \tag{3.81}$$

2. For $2 \leq i \leq N$, in increasing order of i, find $\bar{\alpha}_i$ such that

$$f_i\left(S_0, 0; \sum_{j=1}^{i} \bar{\alpha}_j\right) = \bar{\pi}_i \tag{3.82}$$

3. Set

$$\bar{\alpha} = \sum_{i=1}^{N} \bar{\alpha}_i \tag{3.83}$$

This method constructs $\bar{\alpha}$ by calibrating for each time interval separately, earlier maturities first.

In practice, this method is most often used to build discount curves from bond prices, with instantaneous interest rate $\alpha(t)$. The smoothness of α at knot times $T_1, \ldots, T_{N-1}$ is a major concern.

3.3.2 Non-Parametric Methods

Non-parametric methods do not compute a constant or time-dependent coefficient with finitely many degrees of freedom. Instead, they may:

– Fit a "surface" $\bar{\alpha} = \bar{\alpha}(S_t, t)$ that satisfies the benchmark price constraints. In Sect. 4.2.3 we describe a non-parametric calibration algorithm that produces a volatility surface $\bar{\sigma} = \bar{\sigma}(S_t, t)$.
– Bypass α and directly calibrate an equivalent measure Q such that

$$E_Q(\beta_{T_i} X_i) = \bar{\pi}_i \quad (1 \leq i \leq N) \tag{3.84}$$

In Sect. 4.3 we sketch a Monte-Carlo method that re-weighs path prices after they have been computed with an un-calibrated model. Re-weighing is done without reference to the model that was used to produce the path prices.

In general, the number of degrees of freedom F in non-parametric methods greatly exceeds the number or parameters in the model. In actual computer programs, F can be the number of grid points or the number of Monte-Carlo paths. In addition to benchmark price constraints, other constraints must be used to make the choice of coefficient unique. In this book, worst- and best-case assumptions are used.

4 Scenario-Based Evaluation and Uncertainty

The following problems arise in practice:

- A concrete instance of the selected equity, FX or interest rate model must be chosen, by instantiating its volatility and other coefficients with plausible values. For example, the Black-Scholes model $dS = \mu S_t + \sigma dW$ might be instantiated to $dS = 0.05 S_t + 0.3 dW$.
- Once instantiated, models often prove too weak to represent the market dynamics adequately; in the case of Black-Scholes, this deficiency shows itself in the often cited implied volatility smile.

The second problem can be approached with time- and space-dependency in the volatility and other coefficients. If this implies randomness in the evolution of the volatility, one has created a stochastic volatility model. The first problem does not disappear, however, and some sort of parameter calibration is necessary before the stochastic volatility model can be applied.

Uncertain volatility takes a different approach. Instead of choosing a fixed set of a priori model coefficients, users specify priorities which they would like to see applied when a given portfolio is evaluated under the model. These priorities are initially stated "in prose" and have some economic function. They usually correspond to stochastic control problems and require dynamic programming methods for their solution.

4.1 Preliminaries

Definition 1 (Scenario). *We call a set of (declarative) agent priorities and the (imperative) evaluation rules they imply a scenario.*

Definition 2 (Uncertain coefficients). *Model coefficients which are variable under a given scenario are called* uncertain. *The evaluation rules of the scenario control the instantiation of uncertain coefficients, locally or globally.*

These definitions are not strictly formal. The soundness of the concept needs to be established for each concrete scenario. In this book, we restrict ourselves to two scenarios:

- the worst-case volatility scenario;

34   4 Scenario-Based Evaluation and Uncertainty

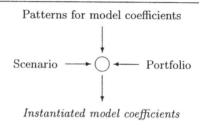

Fig. 4.1. Both scenario and portfolio are required components when model coefficients are instantiated. Model coefficients can, but must not, be restricted by patterns

– the volatility-shock scenario.

We review the worst-case volatility scenario in this chapter. It was first developed by Avellaneda and Parás as the λ-Uncertain Volatility Model or λ-UVM. Algorithmic issues of worst-case scenarios are moved as original work to Part II. The volatility shock scenario is an extension of the worst-case scenario and is discussed, also as original work, in Chapter 9 of Part II.

The benefit of the scenario approach is clear: no definite a-priori choice of model coefficients has to be made. Furthermore, once evaluation rules have been applied to instantiate uncertain coefficients, we're back in the realm of arbitrage pricing theory. On the other hand, as seen in Sect. 3.2, no-arbitrage arguments alone are not sufficient when coefficients are stochastic; disputable assumptions, equilibrium arguments and other methods which are not easily generalizable are required to complete the task.

The scenario approach may yield different instantiations of model coefficients for different portfolios. Figure 4.1 shows how scenario and portfolio are both taken into account when the evaluation rules of the scenario are executed.

The separation into model and scenario is in fact strong enough to reappear in the object-oriented implementation in Part III. Models, scenarios and portfolios all have associated class hierarchies.

In this book, we exclusively focus on the volatility as the only uncertain coefficient. Formally, we assume a filtered probability space $(\Omega, \mathcal{F}, \{\mathcal{F}_t\}, P)$, a one-dimensional Brownian motion W, and some finite time horizon T. In this probability space, let $S = \{S_t\}$ be a security price process with the stochastic differential equation

$$\frac{dS}{S_t} = \mu(S_t, t)\,dt + \sigma(S_t, t)\,dW \tag{4.1}$$

Let $r\colon [0,T] \to \mathbb{R}_+$ be the time-dependent interest rate, and $\beta = \{\beta_t\}$ the corresponding discount process:

$$\beta_t = \exp\left\{-\int_0^t r_s \, ds\right\} \tag{4.2}$$

We assume r and μ are continuous functions that are sufficiently well behaved for our purpose. $\sigma\colon (0,\infty) \times [0,T] \to \mathbb{R}_{++}$ is our uncertain model coefficient.

Definition 3 (Candidate set and scenario measure). *A set*

$$\mathcal{C} \subseteq \{\sigma \mid (4.1) \text{ has a solution}\} \tag{4.3}$$

is called a candidate set *for σ. For each $\sigma \in \mathcal{C}$ there exists a unique measure $Q(\sigma)$ which makes βS a martingale: we say $Q(\sigma)$ is the* scenario measure *for σ.*

Sometimes we also refer to the "scenario σ" or "scenario volatility." The candidate set implements the optional pattern for the uncertain coefficient referred to in Fig. 4.1.

Let the nonnegative, continuous random variable X denote the payoff of a contingent claim at time T. The no-arbitrage price of the contingent claim for fixed σ follows the process

$$F_t(X,\sigma) = \frac{1}{\beta_t} \mathrm{E}_{Q(\sigma)} \left(\beta_T X \mid \mathcal{F}_t\right) \tag{4.4}$$

Extension to portfolios of contingent claims is straightforward. Let $\mathbf{X} = (X_1, \ldots, X_k)^\mathrm{T}$ be a set of $k > 0$ nonnegative contingent claims—a portfolio!— on $(\Omega, \mathcal{F})$, all maturing at time T. (The theory can be easily generalized to contingent claims with different expiration dates.) For any combined position $\lambda = (\lambda_1, \ldots, \lambda_k)^\mathrm{T} \in \mathbb{R}^k$, $\lambda \cdot \mathbf{X}$ is also a—not necessarily nonnegative—random variable on $(\Omega, \mathcal{F})$ and represents the final cashflow at time T for the holder of the portfolio. (At this time we assume that contingent claims are not path-dependent; i.e., their payoff can be written as $g(S_T)$ for some function g. Later, of course, we will include barrier and American options.) The value process $F = \{F_t\}$ is extended to cover combined positions through

$$F_t(\lambda \cdot \mathbf{X}, \sigma) = \sum_{i=1}^k \lambda_i F_t(X_i, \sigma) \tag{4.5}$$

4.2 The Worst-Case Volatility Scenario

We distinguish three concrete worst-case volatility scenarios, or worst-case scenarios for short, each illuminating the exposure to volatility risk from a slightly different perspective. All scenarios have in common that

$$\mathcal{C} = \{\sigma \mid \sigma_{\min} \leq \sigma(S_t, t) \leq \sigma_{\max} \text{ and } (4.1) \text{ has a solution}\} \tag{4.6}$$

36   4 Scenario-Based Evaluation and Uncertainty

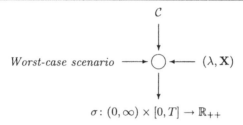

Fig. 4.2. The generic terms of Fig. 4.1 filled in. The worst-case scenario can be tailored to pricing, hedging or calibration situations as described in the text

where $0 < \sigma_{\min} \leq \sigma_{\max}$ represents a prescribed bound. For simplicity, we assume constant bounds, but the theory holds for time-heterogeneous bounds as well. Figure 4.2 illustrates the flow of information that leads from $\mathcal{C}$, $(\lambda, \mathbf{X})$ and the concrete scenario to the selection of $\sigma \in \mathcal{C}$.

The agent priorities in each of the worst-case scenario variations can be informally stated as follows:

Worst-case pricing. Given the portfolio $\mathbf{X}$ and a position $\lambda \in \mathbb{R}^k$ in $\mathbf{X}$. Which $\hat{\sigma} \in \mathcal{C}$ maximizes today's value $F_0(\lambda \cdot \mathbf{X}, \sigma)$?

The optimal hedge-portfolio. Given two portfolios $\mathbf{X}$ and $\bar{\mathbf{X}}$ of resp. k and $\bar{k}$ contingent claims, and a position $\lambda \in \mathbb{R}^k$ in $\mathbf{X}$. For each $\bar{X}_i$, $1 \leq i \leq \bar{k}$, a market price $\bar{\pi}_i$ is known. (Assume, for instance, that the $\bar{X}_i$ are traded frequently, and the X_i are exotic over-the-counter instruments.) Which $\hat{\sigma} \in \mathcal{C}$ maximizes $F_0(\lambda \cdot \mathbf{X}, \sigma)$ under the additional constraint that $F_0(\bar{X}_i, \hat{\sigma}) = \bar{\pi}_i$ for $1 \leq i \leq \bar{k}$?

Calibration. Given a portfolio $\bar{\mathbf{X}}$ of $\bar{k}$ contingent claims, and market prices $\bar{\pi}_i$ for all $\bar{X}_i$, $1 \leq i \leq \bar{k}$. Fix a subjective "prior" $\bar{\sigma} \in \mathcal{C}$. Which $\hat{\sigma} \in \mathcal{C}$ minimizes $\|\sigma - \bar{\sigma}\|$ under the additional constraint that $F_0(\bar{X}_i, \hat{\sigma}) = \bar{\pi}_i$ for each $1 \leq i \leq \bar{k}$? We leave the semantics of the distance $\|\cdot\|$ unspecified.

Section 4.2.1 is dedicated to the the worst-case pricing problem. Section 4.2.2 is a short treatise on the problem of finding the optimal hedge portfolio. Section 4.2.3 investigates calibration issues.

Here and throughout the rest of the work, optimality is denoted by a "^" accent.

4.2.1 Worst-Case Pricing

The objective is to find the volatility coefficient $\hat{\sigma} \in \mathcal{C}$ which maximizes $F_0(\lambda \cdot \mathbf{X}, \sigma)$ for a given vector $\mathbf{X}$ of k contingent claims, and given position $\lambda \in \mathbb{R}^k$. *Sellers* of $\lambda \cdot \mathbf{X}$ are completely hedged against volatility risk within the bounds (4.6) if they charge at least $F_0(\lambda \cdot \mathbf{X}, \hat{\sigma})$. (From this point of view, $\lambda_i > 0$ means X_i is sold, and $\lambda_i < 0$ means X_i is bought. Positive quantities signify liabilities of the seller, while negative quantities signify cash inflow.)

4.2 The Worst-Case Volatility Scenario

The objective must be formalized with care, since $\hat{\sigma}$ may not exist. For instance, assume the final payoff $\lambda \cdot \mathbf{X}$ is convex and continuous, and $\mathcal{C} = \{0.2 - \frac{1}{n} \mid n \geq 6\}$. It is clear that $F_0(\lambda \cdot \mathbf{X}, 0.2 - \frac{1}{n}) \to F_0(\lambda \cdot \mathbf{X}, 0.2)$ from below as $n \to \infty$, yet $0.2 \notin \mathcal{C}$. Nevertheless, $F_0(\lambda \cdot \mathbf{X}, 0.2)$ should be regarded as the worst-case price, and $\sigma = 0.2$ as its scenario coefficient.

Convex Contingent Claims It is instructive to consider the simple case of convex portfolios first. Let $Y = \lambda \cdot \mathbf{X}$, and assume Y can be written $g(S_T(\omega)) = Y(\omega)$ for $\omega \in \Omega$ and some nonnegative convex function $g\colon (0, \infty) \to \mathbb{R}_+$. (For instance, $\mathbf{X}$ might be a vector of European call or put options, with positions $\lambda_i > 0$ throughout). In this case, the Black-Scholes solution is also convex in S. As shown in [52],

Fact 3. *For convex Y, the value process $F(Y, \sigma_{\max})$ is a super-martingale under any measure $Q(\sigma)$ with $\sigma \in \mathcal{C}$. Similarly, the value process $F(Y, \sigma_{\min})$ is a sub-martingale under any measure $Q(\sigma)$ with $\sigma \in \mathcal{C}$. This implies*

$$F_t(Y, \sigma_{\min}) \leq F_t(Y, \sigma) \leq F_t(Y, \sigma_{\max}) \tag{4.7}$$

for $0 \leq t \leq T$ and for all $\sigma \in \mathcal{C}$.

For a nonnegative convex overall position Y, the solution of the maximization problem is thus $\hat{\sigma} = \sigma_{\max}$. Similarly, if Y is negative and concave, $|Y|$ is positive and convex, and $F_t(Y, \sigma) \leq F_t(Y, \sigma_{\min})$ for all $\sigma \in \mathcal{C}$.

General Portfolios Let $Y = \lambda \cdot \mathbf{X}$ be the liability structure at time T for a portfolio $\mathbf{X}$ of k contingent claims and position $\lambda \in \mathbb{R}^k$. This time we make no assumptions about $Y\colon \Omega \to \mathbb{R}$. In [3], Fact 3 is generalized as follows:

Fact 4. *Let $\Sigma\colon \mathbb{R} \to \{\sigma_{\min}, \sigma_{\max}\}$ be the following function:*

$$\Sigma(x) = \begin{cases} \sigma_{\max} & \text{if } x \geq 0 \\ \sigma_{\min} & \text{if } x < 0 \end{cases} \tag{4.8}$$

Given Y, define a value process $\hat{F}(Y) = \{\hat{F}_t(Y)\}$ by $\hat{F}_t(Y) = \hat{f}(S_t, t; Y)$, where $\hat{f}$ is the solution of the partial differential equation

$$\frac{\partial f}{\partial t} + \frac{1}{2} \Sigma\left(\frac{\partial^2 f}{\partial S^2}\right) S_t^2 \frac{\partial^2 f}{\partial S^2} + r_t S_t \frac{\partial f}{\partial S} - r_t f_t = 0 \tag{4.9}$$

with boundary condition $\hat{f}(S_T, T) = Y(S_T)$.

Then $\hat{F}(Y)$ is a super-martingale under any measure $Q(\sigma)$ where $\sigma \in \mathcal{C}$.

The informal rationale is the following: take the original Black-Scholes equation (3.11) and bring $r_t f_t$ to the right side, while observing that the remaining terms on the left side do not contain f. To make f as large as possible, we maximize the only term on the left side which has some degree of freedom: $\frac{1}{2}\sigma S_t^2 \frac{\partial^2 f}{\partial S^2}$. This is accomplished in (4.8).

4 Scenario-Based Evaluation and Uncertainty

Fact 5. Let $\hat{F}(Y)$ be the value process for Y defined in Fact 4. Then

$$\hat{F}_0(Y) = \sup_{\sigma \in \mathcal{C}} F_0(Y, \sigma) \qquad (4.10)$$

Moreover, the σ which yields the supremum is given by (4.8).

Thus, there actually exists a "scenario $\hat{\sigma}$", and it can be constructed locally. Following (3.5), define the $\mathbb{R}^2$-valued replicating strategy $\hat{\theta} = \{\hat{\theta}_t\}$ as

$$\hat{\theta}_t^0 = \beta_t(\hat{F}_t - \frac{\partial}{\partial S}\hat{f}(S_t, t; Y)S_t) \quad \text{and} \quad \hat{\theta}_t^1 = \frac{\partial}{\partial S}\hat{f}(S_t, t; Y) \qquad (4.11)$$

This strategy is termed "super-hedging" in [35] and [36]. It is furthermore observed that (4.11) represents the super-hedging strategy that requires the smallest amount $\hat{F}_0$ of initial funds.

Fact 6. For $c \in \mathbb{R}_{++}$ and two liability structures $Y = \lambda \cdot \mathbf{X}$ and $Z = \lambda' \cdot \mathbf{X}'$,

$$\hat{F}_t(cY) = c\hat{F}_t(Y)$$
$$\hat{F}_t(Y + Z) \leq \hat{F}_t(Y) + \hat{F}_t(Z) \qquad (4.12)$$
$$\hat{F}_t(Y + Z) \geq \hat{F}_t(Y) - \hat{F}_t(-Z)$$

Thus, positions may be scaled, but $\hat{F}$ is nonlinear and sub-additive. (The third statement follows from the second with $F_t(Y) = F_t(Y + Z - Z) \leq F_t(Y + Z) + F_t(-Z)$). Notice also that Fact 6 is valid for $0 \leq t \leq T$, not just for $t = 0$.

4.2.2 The Optimal Hedge Portfolio

Let $\mathbf{X}$ and $\bar{\mathbf{X}}$ be two portfolios of size k and $\bar{k}$, respectively. Assume furthermore that $\lambda \in \mathbb{R}^k$ is a position for X, and $\bar{\pi} \in \mathbb{R}_{++}^{\bar{k}}$ is a market price vector for $\bar{\mathbf{X}}$. ($\mathbf{X}$ might be a book position, and $\bar{\mathbf{X}}$ might be a set of liquid options.) It is a natural restriction to consider only those $\sigma \in \mathcal{C}$ under whose scenario measure $Q(\sigma)$ the prices $\bar{\pi}$ for $\bar{\mathbf{X}}$ are matched. This restriction on $\mathcal{C}$ is defined as follows:

$$\mathcal{C}' = \{\sigma \in \mathcal{C} \mid F_0(\bar{X}_i, \sigma) = \bar{\pi}_i \text{ for } 1 \leq i \leq \bar{k}\} \qquad (4.13)$$

Now let $Y = \lambda \cdot \mathbf{X}$ be the combined payoff of portfolio $\mathbf{X}$. [4] show

Fact 7. Given $\mathbf{X}$, $\bar{\mathbf{X}}$, λ and $\bar{\pi}$. Assume $\hat{\bar{\lambda}} \in \mathbb{R}^{\bar{k}}$ is a finite solution of the following optimization problem in the variables $\bar{\lambda} \in \mathbb{R}^{\bar{k}}$ (the hedging position in the market portfolio) and $\sigma \in \mathcal{C}$:

$$\inf_{\bar{\lambda} \in \mathbb{R}^{\bar{k}}} \left\{ \sup_{\sigma \in \mathcal{C}} F_0(Y + \bar{\lambda} \cdot \bar{\mathbf{X}}, \sigma) - \bar{\lambda} \cdot \bar{\pi} \right\} \qquad (4.14)$$

Let $\hat{\sigma}$ be the scenario volatility for $\hat{\lambda}$ according to Fact 5:

$$F_0(Y + \bar{\lambda} \cdot \bar{\mathbf{X}}, \hat{\sigma}) = \sup_{\sigma \in \mathcal{C}} F_0(Y + \bar{\lambda} \cdot \bar{\mathbf{X}}, \sigma) \qquad (4.15)$$

Then

$$F_0(Y, \hat{\sigma}) = \sup_{\sigma \in \mathcal{C}'} F_0(Y, \sigma) \qquad (4.16)$$

The solution $\hat{\lambda}$ is unique, since the function

$$h(\bar{\lambda}) = \sup_{\sigma \in \mathcal{C}} F_0(Y + \bar{\lambda} \cdot \bar{\mathbf{X}}, \sigma) - \bar{\lambda} \cdot \bar{\pi} \qquad (4.17)$$

is convex and has therefore at most one minimum. Furthermore, under first-order conditions on optimality,

$$\frac{\partial}{\partial \bar{\lambda}_i} \left(F_0(Y + \bar{\lambda} \cdot \bar{\mathbf{X}}, \hat{\sigma}) - \bar{\lambda} \cdot \bar{\pi} \right) \bigg|_{\hat{\lambda}_i} = F_0(\bar{X}_i, \hat{\sigma}) - \bar{\pi}_i = 0 \qquad (4.18)$$

and therefore $F_0(\bar{X}_i, \hat{\sigma}) = \bar{\pi}_i$, for $1 \leq i \leq \bar{k}$.

The position $\hat{\lambda}$ is optimal in the sense that no other position reduces the residual worst-case liability $h(\bar{\lambda})$ by a larger amount. An agent who counterbalances a stake in $\mathbf{X}$ by taking an offsetting position $\hat{\lambda}$ in $\bar{\mathbf{X}}$ needs at most $h(\bar{\lambda})$ additional cash to hedge the combined position, provided the volatility is within the bounds set in $\mathcal{C}$. $\hat{\lambda}$ can thus be regarded as the optimal hedge portfolio under the worst-case scenario.

4.2.3 Calibration to the Worst Case

The goal of calibration is to find an instantiation of the uncertain coefficients that matches observed prices of market instruments exactly. In that sense, the optimal hedge portfolio results from calibrating σ to the market prices $\bar{\pi}$. The method, however, is not satisfactory since it depends on the presence of a book portfolio $\mathbf{X}$. Furthermore, agents cannot introduce subjective prior beliefs about uncertain coefficients; in fact, the resulting scenario σ takes on only extremal values $\sigma_{\min}$ and $\sigma_{\max}$.

For this reason, let us reformulate the problem. Given a portfolio $\bar{\mathbf{X}}$ and a corresponding price vector $\bar{\pi} \in \mathbb{R}^k_{++}$, choose some (constant) prior $\bar{\sigma} \in \mathcal{C}$ that best reflects your subjective beliefs about the volatility of the underlying asset.

For any $\sigma \in \mathcal{C}$ and for any $\omega \in \Omega$, define the distance of σ to $\bar{\sigma}$ on the path $\{S_t(\omega) \mid 0 \leq t \leq T\}$ as

$$d(\sigma, \omega) = \int_0^T \eta\left(\sigma(S_u(\omega), u)^2\right) du \qquad (4.19)$$

where η is a smooth, finite, strictly convex function which attains its minimum at $\bar{\sigma}^2$, i.e. $\eta(\bar{\sigma}^2) = 0$. η is called *pseudo entropy function* and implements a penalty for deviation from the prior—for instance, take $\eta(\sigma^2) = \frac{1}{2}(\sigma^2 - \bar{\sigma}^2)^2$.

With C' as defined in (4.13), Avellaneda *et al* show in [5] that

Fact 8. *Given $\bar{\mathbf{X}}$ and $\bar{\pi}$. Assume $\hat{\lambda} \in \mathbb{R}^{\bar{k}}$ is a finite solution of the following optimization problem in the variables $\bar{\lambda} \in \mathbb{R}^{\bar{k}}$ and $\sigma \in C$:*

$$\inf_{\bar{\lambda} \in \mathbb{R}^{\bar{k}}} \left\{ \sup_{\sigma \in C} F_0(-d(\sigma) + \bar{\lambda} \cdot \bar{\mathbf{X}}, \sigma) - \bar{\lambda} \cdot \bar{\pi} \right\} \tag{4.20}$$

and let $\hat{\sigma} \in C$ be the scenario volatility for $\hat{\lambda}$. Then

$$F_0(-d(\hat{\sigma}), \hat{\sigma}) = \sup_{\sigma \in C'} F_0(-d(\sigma), \sigma) \tag{4.21}$$

In other words, $\hat{\sigma}$ minimizes the penalty. Again, the solution $\hat{\lambda}$ is unique.

Computation of $h(\bar{\lambda})$ In the case of the optimal hedge portfolio, $h(\bar{\lambda})$ is computed by solving (4.9). This approach needs to be modified for calibration.

For fixed η, define the *flux function*

$$\Phi(x) = \sup_{\sigma} \left(\sigma^2 x - \eta(\sigma^2) \right) \tag{4.22}$$

where the supremum is taken over $(\sigma_{\min}, \sigma_{\max})$ and attained at $\sigma = \Phi'(x)$. With $\bar{Y} = \bar{\lambda} \cdot \bar{\mathbf{X}}$ for fixed $\bar{\lambda} \in \mathbb{R}^{\bar{k}}$, define the process $G = \{G_t\}$ as

$$G_t = \sup_{\sigma \in C} F_t(-d(\sigma) + \bar{Y}, \sigma) \tag{4.23}$$

Fact 9. *Given G and $\bar{Y}$. Then $G_t = g(S_t, t)$, where g is the solution of the partial differential equation*

$$\frac{\partial g}{\partial t} + \frac{1}{\beta_t} \Phi\left(\frac{\beta_t}{2} S_t^2 \frac{\partial^2 g}{\partial S^2} \right) + r_t S_t \frac{\partial g}{\partial S} - r_t g_t = 0 \tag{4.24}$$

with boundary condition $g(S_T, T) = \bar{Y}(S_T)$. The supremum in (4.23) is realized at

$$\sigma(S_t, t) = \sqrt{\Phi'\left(\frac{\beta_t}{2} S_t^2 \frac{\partial^2 g}{\partial S^2} \right)} \tag{4.25}$$

By construction, $h(\bar{\lambda}) = G_0$.

The PDE (4.24) can be solved with finite difference methods. Notice that (4.24) is not the pricing equation for $\bar{Y}$; the pricing equation for $\bar{Y}$ is obtained by replacing Φ with $\frac{\Phi'}{2} S_t^2 \frac{\partial^2 g}{\partial S^2}$.

4.3 Minimum-Entropy Calibration

In Sect. 4.2.3 the volatility surface $\sigma(S_t, t)$ of the stochastic model is calibrated to a set of $\bar{k}$ benchmark instruments $\bar{\mathbf{X}}$ with prices $\bar{\pi} \in \mathbb{R}_{++}^{\bar{k}}$. The resulting worst-case risk-neutral measure is a function of $\hat{\sigma}$, i.e. $Q = Q(\hat{\sigma})$.

Although we call this method non-parametric because $\hat{\sigma}$ is constructed node by node on a tree or lattice (if such an implementation is chosen), $\hat{\sigma}$ is still calibrated explicitly.

In this section we describe a method in which no model coefficient is calibrated explicitly, but the worst-case measure is computed directly from the prior measure implicit in the originally selected prior model coefficients. The method generalizes worst-case volatility scenarios.

The following material is taken from [6].

We use the short rate $r = \{r_t\}$ as the underlying and assume it follows the Vasicek model:

$$dr = (\theta - \alpha r)\, dt + \sigma\, dX \qquad (4.26)$$

dX is the random shock, α the speed of mean reversion, and $\frac{\theta}{\alpha}$ the level of mean reversion.

Now assume the process r is sampled N times (in a Monte-Carlo implementation, for example), yielding N paths $\omega_1, \ldots, \omega_N$ of r. The approximate value of any instrument X can then be obtained by computing its discounted expected payoff under these N paths:

$$
\begin{aligned}
F_0^i(X) &\doteq \exp\left(-\int_0^T r_t(\omega_i)\, dt\right) X(\omega_i) \quad (1 \leq i \leq N) \\
F_0(X) &= \frac{1}{N} \sum_{i=1}^N F_0^i(X)
\end{aligned}
\qquad (4.27)
$$

The summation in (4.27) amounts to assigning to each path the weight $\frac{1}{N}$. This uniform probability distribution P of paths is consistent with the *prior model* (4.26). The error made in (4.27) decreases as $N \to \infty$.

Now pick any different probability distribution Q for the paths $\omega_1, \ldots, \omega_N$, i.e. $0 < q_1, \ldots, q_N < 1$ and $\sum_{i=1}^N q_i = 1$. The so-called Kullback-Leibler distance of the new distribution Q to the original, uniform distribution P is

$$
\begin{aligned}
H(Q|P) &= \sum_{i=1}^N Q(\omega_i) \log\left(\frac{Q(\omega_i)}{P(\omega_i)}\right) \\
&= \sum_{i=1}^N Q(\omega_i) \log\left(\frac{Q(\omega_i)}{1/N}\right) \\
&= \log N + \sum_{i=1}^N Q(\omega_i) \log Q(\omega_i) = \log N + \sum_{i=1}^N q_i \log q_i
\end{aligned}
\qquad (4.28)
$$

Here, $0 \leq H(Q|P) \leq \log N$, and $H(Q|P) = 0$ if $Q = P$. Changing the measure from P to Q changes the price of the instrument X:

$$F_0(X \mid Q) \doteq \sum_{i=1}^{N} q_i F_0^i(X) \qquad (4.29)$$

Now let $\bar{\mathbf{X}}$ and $\bar{\pi}$ be a vector of $\bar{k}$ contingent claims and a corresponding price vector, respectively. If N is much greater than $\bar{k}$ it makes sense to ask for the alternative measure Q which correctly prices $\bar{\mathbf{X}}$, given $\bar{\pi}$. A reasonable criterion is to choose Q such that $H(Q|P)$ is minimized. With (4.28), this criterion is equivalent to maximizing the entropy

$$H(Q) = -\sum_{i=1}^{N} q_i \log q_i \qquad (4.30)$$

Under certain assumptions, this constrained entropy optimization problem has a unique solution, which can be found by the method of Lagrange multipliers (see [22], for example). For fixed $\lambda \in \mathbb{R}^{\bar{k}}$, define

$$\begin{aligned} G_0^i(X) &= \exp\left(F_0^i(X)\right) \quad (1 \leq i \leq N) \\ G_0(X) &= \frac{1}{N} \sum_{i=1}^{N} G_0^i(X) \end{aligned} \qquad (4.31)$$

Furthermore define the weights $q_1, \ldots, q_N$ of a measure $Q = Q(\lambda)$ as follows:

$$q_i(\lambda) = \frac{G_0^i(\lambda \cdot \bar{\mathbf{X}})}{G_0(\lambda \cdot \bar{\mathbf{X}})} \quad (1 \leq i \leq N) \qquad (4.32)$$

If the function

$$U_0(\lambda) = \log\left(G_0(\lambda \cdot \bar{\mathbf{X}})\right) - \lambda \cdot \bar{\pi} \qquad (4.33)$$

attains a minimum at $\hat{\lambda}$, then the measure $Q(\hat{\lambda})$ reproduces the prices $\bar{\pi}$ of the benchmark instruments and maximizes $H(Q)$. This can easily be seen by setting the partial derivatives

$$\begin{aligned} \frac{\partial}{\partial \lambda_j} U_0(\lambda) &= \frac{1}{G_0(\lambda \cdot \bar{\mathbf{X}})} \frac{\partial}{\partial \lambda_j} G_0(\lambda \cdot \bar{\mathbf{X}}) - \bar{\pi}_i \\ &= F_0(\lambda_j \bar{X}_j \mid Q(\lambda)) - \bar{\pi}_i \end{aligned} \qquad (4.34)$$

to zero and plugging in $\hat{\lambda}$.

4.4 Scenarios and Nonlinearity

In general, worst-case scenarios lead to nonlinear solutions and are not symmetric for the buy and sell side. Nonlinearity arises because of risk-diversification under mixed convexity of the value of the portfolio. Any position λ in $\mathbf{X}$ has to be priced and hedged as a unit; no "stand-alone" scenario price for X_i can be deduced from $\hat{F}_0$. Sellers of $Y = \lambda \cdot \mathbf{X}$ can hedge against volatility risk within the bounds $\mathcal{C}$ by charging at least $\hat{F}_0(Y)$ and adhering to a "super-hedging" replicating strategy. Vice versa, buyers of Y can hedge against volatility risk within the bounds $\mathcal{C}$ if they pay at most $-\hat{F}_0(-Y)$ and adhere to a "sub-hedging" replicating strategy. The volatility range $[\sigma_{\min}, \sigma_{\max}]$ leads to a corresponding no-arbitrage worst-case price range $[-\hat{F}_0(-Y), \hat{F}_0(Y)]$.

Computationally, nonlinearity requires sophisticated algorithms reduce the combinatorial complexity that arises if the portfolio under consideration contains exotic, path-dependent options. In the remainder of this book, algorithms for barrier and American options are studied in particular.

Part II

Algorithms for Uncertain Volatility Models

5 A Lattice Framework

Nonlinear Black-Scholes models for worst-case scenarios require two kinds of algorithmic techniques:

1. Finite difference methods combined with dynamic programming are used to solve individual PDEs of type (4.9).
2. A collection of PDEs needs to be solved in the right order if exotic options with barrier or American features are involved. Solutions of subordinate PDEs serve as boundary data for PDEs higher up in the hierarchy. (There is only one PDE if the portfolio under consideration contains only vanilla options.)

The sensitivity of the residual portfolio to fluctuations in σ changes if options are taken out through knock-out or early exercise. The so altered portfolio, evaluated independently, may yield an instantiation of σ under the worst-case scenario which differs from the one for the original portfolio. Consequently, it may also yield a worst-case value that differs from the contribution of the remaining options to the worst-case value of the original portfolio, had the option(s) not been taken out. The worst-case value of the reduced portfolio, computed separately, must be used as boundary value for the original portfolio where options are removed by knock-out or early exercise.

An example may help to clarify this explanation. Assume a portfolio of two call options X_1 and X_2 which are identical except for the fact that X_2 allows early exercise, while X_1 does not. The positions are $\lambda_1 = -1$ and $\lambda_2 = 2$, respectively. Let $Y = \lambda \cdot \mathbf{X}$ be the payoff if X_2 is held until maturity, and let $Y' = \lambda_1 X_1$ be the remaining payoff if X_2 is not held until maturity, but exercised early. Figure 5.1 shows the payoff graphically for both cases.

It is clear that the worst-case volatility is $\sigma = \sigma_{\max}$ if X_2 is held until maturity, for Y and $f_t(S_t, t; Y)$ are both convex in S. (Recall that f is the solution of (4.9).) On the other hand, $\sigma = \sigma_{\min}$ from the time on at which X_2 is exercised, for the remainder Y' and thus $f_t(S_t, t; Y')$ are concave in S. In this case, the outlook in terms of exposure to volatility risk is significantly changed. Although the analysis is straightforward in this simple example, the complexity of the problem grows very fast in cases of mixed convexity or exotic options.

48   5 A Lattice Framework

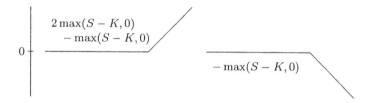

Fig. 5.1. The shape of the final payoff $Y = \lambda \cdot \mathbf{X} = 2X_2 - X_1$ on the left side, and $Y' = \lambda_1 X_1 = -X_1$ on the right side

From now on, worst-case pricing—see Sect. 4.2.1—will be the underlying worst-case scenario. Results are easily applicable to worst-case hedging and calibration.

The current chapter focuses on numerical and general data structure aspects of solvers for PDE's of type (4.9). Chapters 7 and 8 discuss in detail the implications arising from the inclusion of barrier and American options into the portfolio.

5.1 Multi-Lattice Dynamic Programming

The current price of the underlying asset is denoted by $S_0 = s_0$. Let $[s_D, s_U]$ and $[0, T]$ be suitably chosen ranges for the space and time dimensions of the lattice, with $s_D < s_0 < s_U$. Let

$$0 = t_0 < t_1 < \cdots < T_N = T$$

be an equidistant discretization of time, i.e. $t_i = i\,dt$ for $dt = T/N$ and $0 \le i \le N$.

The space dimension need not be uniformly discretized; we will see later that the arbitrary spacing of knock-out barriers requires non-uniformity to avoid slow convergence. Denote the space discretization by

$$s_D = \cdots < s_{-2} < s_{-1} < s_0 < s_1 < s_2 < \cdots = s_U,$$

where for convenience we use the $_D$ and $_U$ subscripts also as numerical index. (In practice, $s_U/s_0 = s_0/s_D \approx 3.5\sqrt{T}$ leads to good results and limits the time complexity in the number of time steps to $O\left(N^{3/2}\right)$. The interested reader is referred to [65].)

5.1.1 Data Structures

We refer to a lattice node by its space and time labels (s_j, t_i), or simply by its space and time indexes (j, i), whichever is more convenient. All PDE's are based on the same discretization. Each PDE, however, is assigned its own lattice instance in memory. Boundary values are shared by copying (and possibly processing) data from one lattice instance to another.

Each lattice instance L is identified by a partial portfolio $\mathbf{X}_L \subseteq \mathbf{X}$ and a position λ_L (which need not be a partial position of λ). If there are only European calls and puts in $\mathbf{X}$, there is only one lattice instance in the computation, identified by top-level $(\mathbf{X}, \lambda)$ (see Chapter 6).

Definition 4 (Lattice signature). *Let L be a lattice instance identified by partial portfolio $\mathbf{X}_L \subseteq \mathbf{X}$ and position λ_L. The pair $(\mathbf{X}_L, \lambda_L)$ is called the signature of L. The size of L is denoted by $|L| = |\mathbf{X}_L| = |\lambda_L|$.*

Often, λ_L is omitted, and only $\mathbf{X}_L$ is used to refer to a lattice instance for simplicity. Lattice instances may be added dynamically during the computation. The set of the signatures of active lattice instances is denoted by $\mathcal{L}$. At all times, $(\mathbf{X}, \lambda) \in \mathcal{L}$.

Definition 5. *Let $L \in \mathcal{L}$ be a lattice instance with signature $(\mathbf{X}_L, \lambda_L)$, and (j, i) a node. $\hat{V}(j, i; L)$ denotes the finite difference approximation of the worst-case value $\hat{F}_i(\lambda_L \cdot \mathbf{X}_L \mid S_i = s_j)$, and $\hat{v}_k(j, i; L)$ denotes its partial derivative in $(\lambda_L)_k$, $1 \leq k \leq |L|$:*

$$\hat{V}(j, i; L) = \hat{F}_i(\lambda_L \cdot \mathbf{X}_L \mid S_i = s_j)$$
$$\hat{v}_k(j, i; L) = \frac{\partial}{\partial (\lambda_L)_k} \hat{F}_i(\lambda_L \cdot \mathbf{X}_L \mid S_i = s_j) \quad (5.1)$$

(Here and thereafter, i and t_i are used interchangingly to index processes such as $\hat{F}$.)

With each node instance $(j, i; L)$ is therefore associated a value/gradient pair

$$\left[\hat{V}(j, i; L), \quad (\hat{v}_k(j, i; L) \mid 1 \leq k \leq |L|) \right]$$

that is stored in the lattice instance's private memory. If it is clear which lattice instance L is denoting, or if L is not significant, L is omitted.

Not all value/gradient pairs need to be accessible at the same time. Two general rules must to be observed, however:

Internal consistency: For the finite difference scheme to work, all time $i + 1$ value/gradient node instances need to be available when time i node instances are computed.

50   5 A Lattice Framework

External consistency: A node instance $(j, i; L')$ needs to be available if the computation of node instance $(j, i; L)$, $\mathbf{X}_{L'} \subseteq \mathbf{X}_L$, requires the lookup of a boundary value associated with partial portfolio/position $(\mathbf{X}_{L'}, \lambda_{L'})$.

The second rule motivates the general policy to process an existing lattice instance L' before any lattice instance L if $|L'| < |L|$. Furthermore, a mechanism must be implemented which automatically inserts a new lattice instance with an appropriate signature into $\mathcal{L}$ whenever the second rule is being violated (exception handling).

5.1.2 Dataflow for Explicit Methods

Definition 6. *We say that the node instance $(j, i; L)$ belongs to the continuation region if no L'-lookup is necessary to determine the worst-case value for it, for any lattice instance $L' \neq L$.*

Figure 5.2 shows the dataflow for an explicit forward Euler one-level scheme for a PDE of type (4.9) within the continuation region.

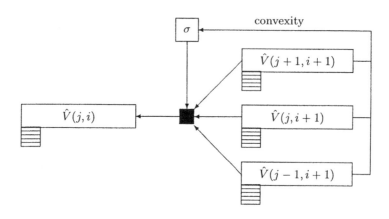

Fig. 5.2. Dataflow for explicit one-level finite differencing in the continuation region. Values at time $i+1$ nodes are first used to compute the worst-case volatility. The black box signifies the finite difference approximation for the PDE. The compartmentalized node attachments symbolize the gradient $(\hat{v}_k(\cdot, \cdot))_k$

If a node instance $(j, i; L)$ turns out to require boundary data from $L' \neq L$, the scheme in Fig. 5.2 may or may not be bypassed, depending on whether $\hat{V}(j, i; L)$ can be determined unconditionally (knock-out) or not (agent's choice like early exercise).

Notice that data flows from time $i+1$ to time i slices for both instantiation of the uncertain coefficient and actual rollback.

5.1.3 Dataflow for Mixed Explicit/Implicit Methods

Mixed explicit/implicit methods such as Crank-Nicholson introduce a lag of one time slice between the instantiation of the uncertain coefficient and the actual rollback, as shown pictorially in Fig. 5.3.

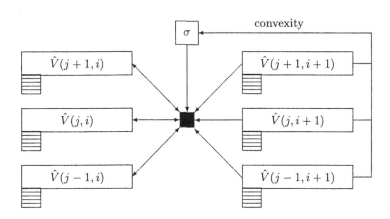

Fig. 5.3. Dataflow for mixed explicit/implicit one-level finite differencing in the continuation region. Values at time $i+1$ nodes are first used in an explicit fashion to compute the worst-case volatility at all space levels. The black box represents one equation in the linear system of equations, instantiated with the local worst-case volatility. The bi-directional arrows on the left side indicate the implicit nature of the system

This lag is necessary to preserve the simplicity of the tridiagonal system of linear equations associated with the rollback step from time $i+1$ to time i. The nonlinearity introduced by the worst-case scenario is taken care of entirely in the explicit instantiation of σ.

Mixed methods cause more problems if the transfer of boundary values between lattice instances is not unconditional. Iterative refinement methods such as SOR must then be employed since the replacement of one $\hat{V}(j,i;L)$ affects all other $\hat{V}(\cdot,i;L)$'s, through their implicit connection.

5.2 Numerical Issues

It is common practice to solve PDE's of the Black-Scholes type using a lattice whose space dimension is discretized uniformly on a logarithmic scaling. It is also desirable for improved accuracy that knock-out barriers coincide with spatial levels of the lattice whenever possible. Since a) the number of distinct barriers in the portfolio is not limited, b) all instruments and thus all barriers must be watched simultaneously under worst-case scenarios, and c) uniform spacing can match at most one barrier and s_0, or two barriers at the same time ([70] and [18]), it is reasonable to modify the standard procedure to allow for non-uniformity.

To guarantee stability explicit forward Euler schemes require the von Neumann condition $dt/(\Delta x)^2 \leq 1/2$ to hold. Here, Δx is the spatial step size after a suitable variable transformation. Equivalently, one may require the transition weights assigned to the arrows in Fig. 5.2 to remain positive (see [75] or [78]).

We present an algorithm that matches all barriers except those that are *very* close together, retains uniform spacing between barriers, and obeys the von Neumann stability condition for explicit schemes (this condition is lifted for Crank-Nicholson).

The following exposition follows [2].

Let the factors $U_j = s_{j+1}/s_j$ resp. $D_j = s_{j-1}/s_j$ represent the size of the up resp. down moves at each spatial level. Instead of using the increments U_j and D_j directly, however, we switch to their logarithms and work with quantities $\overline{\sigma}_U^j$ and $\overline{\sigma}_D^j$ satisfying

$$U_j = 1/D_{j+1} = e^{\overline{\sigma}_U^j \sqrt{dt}}$$
$$D_j = 1/U_{j-1} = e^{-\overline{\sigma}_D^j \sqrt{dt}} \tag{5.2}$$

for $D \leq j \leq U$. dt is the time increment determined from an initial target increment $dt_{\max}$.

Equation (4.9) is formulated with the riskneutral drift r_t. We generalize and write $\mu_t = r_t - d_r$ instead, where d_t denotes a dividend rate, foreign interest rate or storage cost, depending on the properties of the underlying asset. It is assumed that lower and upper bounds

$$\mu_{\min} \leq \mu_t \leq \mu_{\max} \qquad (0 \leq t \leq T) \tag{5.3}$$

are known.

For simplicity, we assume that there are n up-and-out barriers

$$s_0 < b_1 < b_2 < \cdots < b_n < \infty,$$

and no down-and-out barriers. Extension to down-and-out barriers in both algorithms and propositions is straightforward. By convention, $s_0 = b_0$ is also treated as a barrier.

Proposition 1. *Given barriers $b_0, \ldots, b_n$, a target time step $dt_{\max}$, volatility bounds $\sigma_{\min}$, $\sigma_{\max}$, and drift bounds $\mu_{\min}$, $\mu_{\max}$. If the algorithm in Fig. 5.4 is used to compute spatial increments $\overline{\sigma}_U^j$, $\overline{\sigma}_D^j$ for $D \leq j \leq U$ together with*

Input: barriers $b_0, \ldots, b_n$, $dt_{\max}$, $\sigma_{\min}$, $\sigma_{\max}$, $\mu_{\min}$, $\mu_{\max}$
Output: dt, $\overline{\sigma}_U^j$, $\overline{\sigma}_D^j$ for $0 \leq j \leq U$
(extension to cover down-and-out barriers as well is straightforward)

1. Set $\overline{\mu} := \max\{|\mu_{\min}|, |\mu_{\max}|\}$
2. Set $dt := dt_{\max}$. This is the initial guess, to be adjusted later
3. Repeat for $i = 0, \ldots, n$:
    a) Set $\overline{\sigma} := 2\sigma_{\max}$ (see remark in text)
    b) If $i = n$ then skip the next step (there are no more barriers above b_n)
    c) Increase $\overline{\sigma}$ such that $b_i e^{k\overline{\sigma}\sqrt{dt}} = b_{i+1}$ for some $k \in \mathbb{N}$. If no such k exists (i.e., $\ln(b_{j+1}/b_j) < \overline{\sigma}\sqrt{dt}$), abort and report an error (see remark in text)
    d) Set
    $$dt' := \left[\frac{2\sigma_{\min}^2}{\overline{\sigma}(\sigma_{\max}^2 + 2\overline{\mu})}\right]^2$$
    Check if $dt < dt'$. If yes, skip the next step (dt has passed the test)
    e) dt is too big: choose a new $dt > 0$ such that $dt < dt'$ and start over with step 3 (for instance, set $dt = 0.9\, dt'$)
    f) For all s_j such that $b_i < s_j < b_{i+1}$ (or simply $b_i < s_j$ if $i = n$), set $\overline{\sigma}_U^j := \overline{\sigma}_D^j := \overline{\sigma}$. In addition, set $\overline{\sigma}_U^{j_0} := \overline{\sigma}$ where $s_{j_0} = b_i$, and if $i < n$, set $\overline{\sigma}_D^{j_1} := \overline{\sigma}$ where $s_{j_1} = b_{i+1}$

Fig. 5.4. Discretizing space while preserving the von Neumann condition. The input $dt_{\max}$ indicates the desirable time step, from which spatial increments $\overline{\sigma}_U^j$ and $\overline{\sigma}_D^j$ are derived. The output dt is equal to $dt_{\max}$ if no adjustments are necessary, smaller otherwise (see step 3e). The algorithm matches one barrier at a time, starting with $b_0 = s_0$.

a possible adjustment of $dt_{\max}$ to dt, then the explicit forward Euler approximation of (4.9) shown in Fig. 5.5 obeys the von Neumann stability condition. In particular, the variables P_U and P_D satisfy

$$P_U, P_D > 0$$
$$P_U + P_D < \frac{1}{2} \qquad (5.4)$$

Furthermore, the barriers $b_0, b_1, \ldots, b_n$ are all matched if the algorithm in Fig. 5.5 terminates successfully (no error is triggered in step 3c).

Proof. See [2] for a full proof. Here, let us only apply the transformation $X = \log(S)$ to (4.9) to get

$$\frac{\partial f}{\partial t} + \frac{1}{2}\Sigma^2\left\{e^{-2X}\left(\frac{\partial^2 f}{\partial X^2} - \frac{\partial f}{\partial X}\right)\right\}\left(\frac{\partial^2 f}{\partial X^2} - \frac{\partial f}{\partial X}\right) + \mu_t\frac{\partial f}{\partial X} - r_t f_t = 0 \quad (5.5)$$

with

Input: Lattice instance L, time $i\,dt$, $\overline{\sigma}_U^j$, $\overline{\sigma}_D^j$ for $D \leq j \leq U$
Output: $\hat{V}(j,i;L)$ for $D \leq j \leq U$

1. Repeat for $D < j < U$:
   a) Define

   $$P_U(\sigma) = \frac{\sigma^2}{\overline{\sigma}_U^j \overline{\sigma}_D^j + (\overline{\sigma}_U^j)^2}\left(1 - \frac{\overline{\sigma}_D^j \sqrt{dt}}{2}\right) + \frac{\mu \overline{\sigma}_D^j \sqrt{dt}}{\overline{\sigma}_U^j \overline{\sigma}_D^j + (\overline{\sigma}_U^j)^2}$$

   $$P_D(\sigma) = \frac{\sigma^2}{\overline{\sigma}_U^j \overline{\sigma}_D^j + (\overline{\sigma}_D^j)^2}\left(1 + \frac{\overline{\sigma}_U^j \sqrt{dt}}{2}\right) - \frac{\mu \overline{\sigma}_U^j \sqrt{dt}}{\overline{\sigma}_U^j \overline{\sigma}_D^j + (\overline{\sigma}_D^j)^2}$$

   $$P_M(\sigma) = 1 - P_U(\sigma) - P_D(\sigma)$$

   b) Set

   $$\hat{V}(j,i;L) := e^{-r t_i dt} \max_\sigma \Big\{ P_U(\sigma)\,\hat{V}(j+1, i+1; L) \\ + P_M(\sigma)\,\hat{V}(j, i+1; L) + P_D(\sigma)\,\hat{V}(j-1, i+1; L) \Big\}$$

   where the maximum is taken over $\{\sigma_{\min}, \sigma_{\max}\}$

2. Extrapolate to get $\hat{V}(D,i;L)$ and $\hat{V}(U,i;L)$

Fig. 5.5. The explicit forward Euler scheme to compute the worst-case value $\hat{V}(j,i;L)$ at all spatial levels $s_D, \ldots, s_U$ from the $\hat{V}(\cdot, i+1; L)$. The gradient is computed similarly. This algorithm corresponds to Fig. 5.2

$$\Sigma^2\{C\} = \begin{cases} \sigma_{\max}^2 & \text{if } C \geq 0 \\ \sigma_{\min}^2 & \text{if } C < 0 \end{cases} \tag{5.6}$$

The explicit finite difference approximations for (5.5) are as follows. For the time axis, the forward difference

$$\frac{\partial f}{\partial t} \doteq \frac{\hat{V}(j, i+1) - \hat{V}(j,i)}{dt} \tag{5.7}$$

is used. On the space axis, centered differences for both the first and second partial derivatives are used. Since the upward and downward displacement might differ, the formulas are slightly more complex than usual:

$$\frac{\partial f}{\partial X} \doteq \frac{1}{\alpha \sqrt{dt}} \left[\left(\overline{\sigma}_D^j\right)^2 \hat{V}(j+1, i+1) \right.$$
$$\left. - \left(\overline{\sigma}_U^j\right)^2 \hat{V}(j-1, i+1) - \left(\left(\overline{\sigma}_D^j\right)^2 - \left(\overline{\sigma}_U^j\right)^2\right) \hat{V}(j, i+1) \right] \tag{5.8}$$

$$\frac{\partial^2 f}{\partial X^2} \doteq \frac{2}{\alpha \, dt} \left[\overline{\sigma}_D^j \hat{V}(j+1, i+1) \right.$$
$$\left. + \overline{\sigma}_U^j \hat{V}(j-1, i+1) - (\overline{\sigma}_D^j + \overline{\sigma}_U^j) \hat{V}(j, i+1) \right]$$

where

$$\alpha = \overline{\sigma}_U^j \left(\overline{\sigma}_D^j\right)^2 + \overline{\sigma}_D^j \left(\overline{\sigma}_U^j\right)^2 \tag{5.9}$$

Algebra shows that the weights P_U, P_M and P_D computed in the algorithm in Fig. 5.4 replicate the approximation (5.7) and (5.8). They furthermore satisfy (5.4) by construction: crucial is step 3d.

The barriers are matched by construction as well. □

P_U, P_D and $P_M = 1 - P_U - P_D$ can be regarded as probabilities. The property $P_U + P_D < \frac{1}{2}$ guarantees that the middle weight is always at least $\frac{1}{2}$; this has been found empirically to lead to a significant improvement in accuracy (a small P_M effectively turns the explicit scheme into a binomial tree method).

Note that the algorithm in Fig. 5.5 matches the barriers regardless of the validity of the von Neumann condition. The algorithm can thus be used unmodified for mixed explicit/implicit schemes (and indeed is). The algorithm in Fig. 5.4 can be significantly simplified in the mixed case (the test with dt' can be omitted).

Two further remarks should be made. First, step 3a in the algorithm in Fig. 5.5 can safely be replaced by

3a' Set $\overline{\sigma} := \sqrt{2} \, \sigma_{\max}$

In this case $P_U + P_D < \frac{1}{2}$ only if $\overline{\sigma}_U^j = \overline{\sigma}_D^j$. For $\overline{\sigma}_U^j \neq \overline{\sigma}_D^j$, the upper bound becomes $P_U + P_D < 1$ instead. This still guarantees $P_M > 0$ and therefore does not break the probability framework of the derivation. Moreover, $\overline{\sigma}_U^j \neq \overline{\sigma}_D^j$ for at most n spatial levels of the lattice (n is the number of barriers). The ratio of the number of "good" j's ($\overline{\sigma}_U^j = \overline{\sigma}_D^j$) over the number of "bad" j's ($\overline{\sigma}_U^j \neq \overline{\sigma}_D^j$) is therefore negligible as the granularity of the lattice gets finer.

Moreover, the algorithm may trigger an error in step 3c. If two barriers are too close to each other, one of them must be ignored and the algorithm is restarted with the number of barriers reduced by one.

6 Algorithms for Vanilla Options

Vanilla options are standard European calls and puts on the underlying asset. Before we discuss algorithms for barrier and American options we illustrate uncertain volatility for portfolios of vanilla options with an example.

Let $\mathbf{X}$ be a portfolio of k European call or put options $X_1, \ldots, X_k$ with positions $\lambda_1, \ldots, \lambda_k$. The restriction to vanilla options simplifies pricing under uncertain volatility scenarios as follows:

- No subordinate PDE's need to be solved to feed boundary data. In the lattice framework of Chapter 5, only the top-level lattice instance $(\mathbf{X}, \lambda)$ is needed.
- There are no numerical issues with respect to barriers such as those discussed in Sect. 5.2.

The single lattice instance is built based on the algorithm in Fig. 5.4, simplified as shown in Fig. 6.1 (recall that D and U are the number of lattice levels below and above the mid-level, respectively). Once the lattice instance is built, the options portfolio is priced on it. The explicit Euler scheme of Fig. 5.5 is applicable.

Now consider a butterfly spread of four European call options as shown in Fig. 6.2. All four options mature in $T = 60$ days.

The butterfly spread is evaluated under four volatility scenarios:

1. A worst-case volatility scenario for the seller of the portfolio with $\sigma_{\min} = 0.1$ and $\sigma_{\max} = 0.3$.
2. A linear-volatility scenario with constant $\sigma = 0.1$.
3. A linear-volatility scenario with constant $\sigma = 0.3$.
4. A worst-case volatility scenario for the buyer of the portfolio with $\sigma_{\min} = 0.1$ and $\sigma_{\max} = 0.3$.

The worst-case scenario for the seller has been defined in Sect. 4.2.1: if $\hat{\sigma} \in \mathcal{C}$ is the volatility surface that maximizes $F_0(\lambda \cdot \mathbf{X}, \sigma)$, then sellers of the position λ in $\mathbf{X}$ need only charge $\hat{F}_0(\lambda \cdot \mathbf{X}) = F_0(\lambda \cdot \mathbf{X}, \hat{\sigma})$ in order to be able to super-replicate the portfolio without additional cash inflow. A similar point of view can be taken by buyers of position λ in $\mathbf{X}$: with initial funds in the amount

$$- \sup_{\sigma \in \mathcal{C}} F_0(-\lambda \cdot \mathbf{X}, \sigma) \qquad (6.1)$$

Input: $dt_{\max}$, $\sigma_{\min}$, $\sigma_{\max}$, $\mu_{\min}$, $\mu_{\max}$
Output: dt, $\overline{\sigma}_U^j$, $\overline{\sigma}_D^j$ for $D \leq j \leq U$

1. Set $\overline{\mu} := \max\{|\mu_{\min}|, |\mu_{\max}|\}$
2. Set $\overline{\sigma} := 2\sigma_{\max}$
3. Set

$$dt' := \left[\frac{2\sigma_{\min}^2}{\overline{\sigma}(\sigma_{\max}^2 + 2\overline{\mu})}\right]^2$$

4. Set

$$dt := \min\left(dt_{\max}, 0.9\, dt'\right)$$

5. For all lattice levels $D \leq j \leq U$ set $\overline{\sigma}_U^j := \overline{\sigma}_D^j := \overline{\sigma}$

Fig. 6.1. Discretizing space for a portfolio of vanilla options. The input $dt_{\max}$ indicates the desirable time step, from which spatial increments $\overline{\sigma}_U^j$ and $\overline{\sigma}_D^j$ are derived. The output dt is equal to $dt_{\max}$ if no adjustments are necessary, smaller otherwise (in step 4)

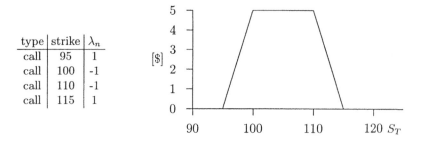

Fig. 6.2. A butterfly spread consisting of four call options, all maturing in $T = 60$ days. The graph on the right side shows the payoff at maturity

the portfolio can be delta-hedged without the need for additional cash inflow, assuming the volatility bounds are not violated.

The worst-case scenarios for the seller versus the buyer's point of view lead to different worst-case volatility functions $\hat{\sigma}\colon (S_t, t) \to \mathbb{R}_{++}$. In this book worst-case scenarios are almost always evaluated from the seller's perspective. We have included the fourth scenario above which takes on the buyer's perspective for illustration. We leave it to the reader to jusify (6.1).

The graph in Fig. 6.3 shows the prices of the butterfly spread for different values of S_0 and interest rate $r = 0.03$, under all four volatility scenarios.

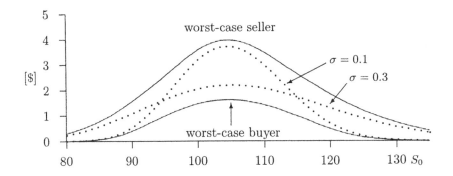

Fig. 6.3. The butterfly spread is evaluated under four volatility scenarios. For both worst-case scenarios, $\sigma_{\min} = 0.1$ and $\sigma_{\max} = 0.3$

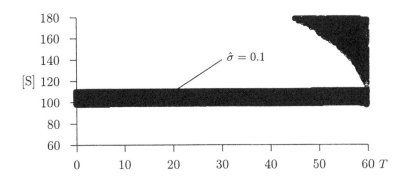

Fig. 6.4. The worst-case volatility surface for the seller, for a butterfly spread that matures in $T = 60$ days. Black regions represent $\hat{\sigma}(S_t, t) = 0.1$, white regions indicate $\hat{\sigma}(S_t, t) = 0.3$. This graph was generated with the forward Euler scheme in Fig. 5.5

The linear prices for constant volatility are always between the worst-case prices, demonstrating that neither a constant volatility coefficient $\sigma = 0.1$ nor a constant coefficient $\sigma = 0.3$ are solutions of (4.10) or (6.1).

The graph in Fig. 6.4 shows the volatility function $\hat{\sigma} \colon (S_t, t) \to \mathbb{R}_{++}$ (we also say "volatility surface") for the seller's worst-case volatility scenario. Black regions indicate lattice nodes where the algorithm in Fig. 5.5 sets $\hat{\sigma}(S_t, t) = 0.1$. White regions indicate lattice nodes where the algorithm in Fig. 5.5 sets $\hat{\sigma}(S_t, t) = 0.3$. The surface is consistent with Fig. 6.3: low volatility around $S = 105$ increases the value of the portfolio.

The upper-right area of the graph has no gamma close to maturity, because the butterfly spread has no payoff for $S_T \geq 115$. The graph shows numerical noise in areas into which the explicit backward induction has not propagated curvature yet, since it does not matter whether $\sigma_{\min}$ or $\sigma_{\max}$ are chosen if $\hat{V}(j+1, i+1; L)$, $\hat{V}(j, i+1, L)$, and $\hat{V}(j-1, i+1; L)$ of Fig. 5.5 (step 1b) lie on a straight line.

7 Algorithms for Barrier Options

We consider barrier options that knock out the first time the price S_t of the underlying asset crosses a predetermined knock-out barrier b. This is one flavor of barrier options; a timing feature is added in [29], where options loose a fraction of their value for every day they spend above (or below) b.

Consider a portfolio $\mathbf{X}$ with position λ_1 in barrier option X_1 with knock-out barrier $b > s_0$ and positions $\lambda_2, \ldots, \lambda_k$ in $k-1$ vanilla options $X_2, \ldots, X_k$. Let all options mature at time T. The payoff at time T is path-dependent: depending on whether the underlying asset has reached the barrier b in the time interval $[0, T]$ or not, the owner of the portfolio receives $\sum_{i=2}^{k} \lambda_i X_i$ or $\lambda \cdot \mathbf{X}$, respectively.

The situation is shown pictorially in Fig. 7.1. Path 1 crosses the barrier u at time t, path 2 doesn't. When path 1 hits the barrier, X_1 becomes worthless. As the portfolio is reduced by one instrument, its sensitivity to volatility fluctuations between times t and T is likely to differ from the sensitivity of the original, unaltered portfolio $(\mathbf{X}, \lambda)$. The worst-case volatility from time t onwards is therefore likely to be different for the partial versus the original portfolio. Hence, two instances of the worst-case pricing problem must be solved, for $(\mathbf{X}, \lambda)$ and for $(\mathbf{Y}, \lambda')$, respectively, where $\mathbf{Y} = (X_2, \ldots, X_k)^{\mathrm{T}}$ and $\lambda' = (\lambda_2, \ldots, \lambda_k)^{\mathrm{T}}$.

Two worst-case pricing problems correspond to two lattice instances L_1 and L_2, each assigned to solve a PDE of type (4.9). The boundary conditions imposed on the two PDE's, however, differ. L_2 is used to solve an initial-value problem with initial value

$$\hat{f}(S_T, T; \lambda', \mathbf{Y}) = \lambda' \cdot \mathbf{Y}(S_T) \tag{7.1}$$

as the partial portfolio $(\mathbf{Y}, \lambda')$ contains only vanilla options. L_1 is used to solve an initial-boundary-value problem with initial value

$$\hat{f}(S_T, T; \lambda, \mathbf{X}) = \lambda \cdot \mathbf{X}(S_T) \tag{7.2}$$

and boundary value

$$\hat{f}(u, t; \lambda, \mathbf{X}) = \hat{f}(u, t; \lambda', \mathbf{Y}) \tag{7.3}$$

for $0 \leq t \leq T$. Under the assumption that L_1 and L_2 match the barrier u at level j_u, (7.3) is reflected within the finite difference framework by the identity of values

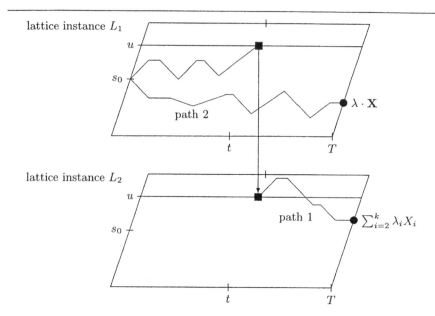

Fig. 7.1. Two paths taken by the underlying asset. Path 1 crosses the barrier u and looses the position in X_1 at time t, consequently shifting to lattice instance L_2 whose signature does not contain X_1. Path 2 stays below the barrier and leaves the portfolio intact until expiration

$$\hat{V}(j_u, i; L_1) = \hat{V}(j_u, i; L_2) + 0 \qquad (7.4)$$

and the identity of gradients

$$\begin{aligned}
\hat{v}_1(j_u, i; L_1) &= 0 \\
\hat{v}_2(j_u, i; L_1) &= \hat{v}_1(j_u, i; L_2) \\
&\cdots \\
\hat{v}_k(j_u, i; L_1) &= \hat{v}_{k-1}(j_u, i; L_2)
\end{aligned} \qquad (7.5)$$

for all time slices $0 \leq i \leq N$. We write "+ 0" in (7.4) to indicate that any potential rebate paid as a result of X_1's knock-out must be added to the residual worst-case liability $\hat{V}(j_u, i; L_2)$. The identities (7.4) and (7.5) replace the transactions shown in Figs. 5.2 and 5.3. External consistency requires furthermore that nodes of L_2 are always processed before corresponding nodes of L_1.

A remark regarding path-dependency The candidate set $\mathcal{C}$ defined in Def. 3 contains only path-*independent* elements $\sigma \colon (\mathbb{R}_{++} \times [0, T]) \to \mathbb{R}_{++}$. To solve separate pricing problems on distinct lattice instances and transfer boundary data makes the volatility path-dependent and thus leads to a worst-case volatility scenario that may not be part of $\mathcal{C}$. Without proof, however, we point out that the worst-case volatility is path-independent (i.e., recombining on a discrete lattice) for paths that remain within a single lattice instance. Only where paths hit boundaries and a jump between lattice instances occurs may volatilities diverge. Each realized path can experience only a finite number of such jumps.

We refrain from changing Def. 3 formally, but ask the reader to keep this remark in mind. In any case, the subsequent definitions that contain terms such as $\sup_{\sigma \in \mathcal{C}}(E_{Q(\sigma)}(\ldots))$ remain consistent under either interpretation of $\mathcal{C}$, since jumps are always explicitly expressed recursively through terms that use different σ' and σ depending on whether the jump occurs or not.

7.1 The Hierarchy of PDE's

We have seen that two lattice instances have to be created if the portfolio contains one barrier option, thus doubling the cost of solving the worst-case pricing problem. The immediate question is: what is the number of lattice instances in the general case, and what are their signatures? How expensive is it to compute worst-case values for portfolios that contain more than one barrier option?

We answer this question for any portfolio that contains up-and-out, down-and-out and double-barrier knock-out options. Up-and-out barrier options knock out if the asset price reaches a barrier $u > s_0$, as in the example in Fig. 7.1. Down-and-out barrier options knock out if the asset prices falls to a level $d < s_0$. Double-barrier options knock out as soon as the asset reaches a barrier $u > s_0$, or falls to a level $d < s_0$: the interval $[d, u]$ defines a corridor in which the double-barrier option is alive.

The following sections closely follow [2].

7.1.1 Construction

Let each instrument of the portfolio $\mathbf{X}$ be associated with an up-and-out barrier $u(X_i)$ and a down-and-out barrier $d(X_i)$, $1 \leq i \leq k$. Vanilla options are modelled by setting $d(X_i) = 0$ and $u(X_i)$ to a very large number (preferably larger than s_U, the upper boundary of the finite difference lattice). For a single up-and-out barrier option with barrier b, $d(X_i) = 0$ and $u(X_i) = b$. For a single down-and-out barrier option with barrier b, $d(X_i) = b$ and $u(X_i)$ very large. For double barrier options, $d(X_i)$ and $u(X_i)$ are both set to the respective barriers.

The open asset-price interval in which the instrument X_i is possibly alive is denoted by $a(X_i) = (d(X_i), u(X_i))$, $1 \leq i \leq k$. Let $\mathbf{Y} \subseteq \mathbf{X}$ be a partial portfolio with $k' \leq k$ instruments. Define

$$A(\mathbf{Y}) = \bigcap_{i=1}^{k'} a(Y_i) \tag{7.6}$$

$A(\mathbf{Y})$ is also open. Let

$$\begin{aligned} U(\mathbf{Y}) &= \sup A(\mathbf{Y}) = \min_{i=1}^{k'} u(Y_i) \\ D(\mathbf{Y}) &= \inf A(\mathbf{Y}) = \max_{i=1}^{k'} d(Y_i) \end{aligned} \tag{7.7}$$

$[D(\mathbf{Y}), U(\mathbf{Y})]$ is the closure of $A(\mathbf{Y})$. $U(\mathbf{Y})$ is the smallest up-and-out barrier of the instruments in $\mathbf{Y}$. Similarly, $D(\mathbf{Y})$ is the largest down-and-out-barrier in $\mathbf{Y}$. If the underlying asset stays within $A(\mathbf{Y})$, an initial position in $\mathbf{Y}$ will remain intact until expiration.

Definition 7 (Extensions). *Let $\mathbf{Y} \subseteq \mathbf{X}$ be a partial portfolio with $k' \leq k$ instruments.*

$$B_U(\mathbf{Y}) = \{1 \leq i \leq k' \mid u(Y_i) > U(\mathbf{Y})\} \tag{7.8}$$

is called the upper extension *of $\mathbf{Y}$. Correspondingly,*

$$B_D(\mathbf{Y}) = \{1 \leq i \leq k' \mid d(Y_i) < D(\mathbf{Y})\} \tag{7.9}$$

is called the lower extension *of $\mathbf{Y}$. The vectorized versions of B_U and B_D are*

$$\begin{aligned} \mathbf{B}_U(\mathbf{Y}) &= \text{select}\,(\mathbf{Y}, B_U(\mathbf{Y})) \\ \mathbf{B}_D(\mathbf{Y}) &= \text{select}\,(\mathbf{Y}, B_D(\mathbf{Y})) \end{aligned} \tag{7.10}$$

Similarly, a position λ' in $\mathbf{Y}$ is reduced to

$$\begin{aligned} \lambda'_U(\mathbf{Y}) &= \text{select}\,(\lambda', B_U(\mathbf{Y})) \\ \lambda'_D(\mathbf{Y}) &= \text{select}\,(\lambda', B_D(\mathbf{Y})) \end{aligned} \tag{7.11}$$

$B_U(\mathbf{Y})$ resp. $B_D(\mathbf{Y})$ indicate which instruments in $\mathbf{Y}$ remain possibly alive when the price of the underlying asset crosses $U(\mathbf{Y})$ resp. $D(\mathbf{Y})$. $B_U(\mathbf{Y})$ and $B_D(\mathbf{Y})$ are sets; the corresponding partial portfolios $\mathbf{B}_U(\mathbf{Y})$ and $\mathbf{B}_D(\mathbf{Y})$ are possibly empty. $(\mathbf{B}_U(\mathbf{Y}), \lambda'_U(\mathbf{Y}))$ and $(\mathbf{B}_D(\mathbf{Y}), \lambda'_D(\mathbf{Y}))$ are the signatures of the lattice instances that feed the boundary data at $U(\mathbf{Y})$ and $D(\mathbf{Y})$. For empty $\mathbf{B}_U(\mathbf{Y})$ or $\mathbf{B}_D(\mathbf{Y})$, no lookup is necessary.

If $U(\mathbf{Y})$ is very large (as is the case if $\mathbf{Y}$ consists of vanilla options only), it will lie outside the finite lattice. Similarly, $D(\mathbf{Y}) = 0$ also falls outside the lattice. In these cases, no additional lattice instances need to be maintained.

Figure 7.2 gives an example for $k = 4$. The sequences of up-and-out and down-and-out barriers are

$$\begin{aligned} s_0 &< u(X_4) < u(X_1) = u(X_3) < u(X_2) \\ s_0 &> d(X_1) = d(X_3) > d(X_2) = d(X_4) \end{aligned} \tag{7.12}$$

7.1 The Hierarchy of PDE's 65

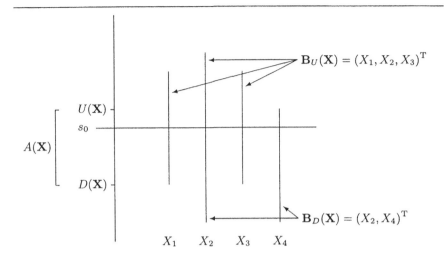

Fig. 7.2. A portfolio $\mathbf{X} = (X_1, X_2, X_3, X_4)^T$ consisting of $k = 4$ options and its upper and lower extensions, $\mathbf{B}_U(\mathbf{X})$ and $\mathbf{B}_D(\mathbf{X})$. The vertical axis marks the price of the underlying asset. $U(\mathbf{X})$ is the smallest up-and-out barrier among the up-and-out barriers in $\mathbf{X}$. Similarly, $D(\mathbf{X})$ is the smallest down-and-out barrier among the down-and-out barriers in $\mathbf{X}$. Each option X_i is represented by a vertical bar whose endpoints are its barriers $u(X_i)$ and $d(X_i)$

Partial portfolio	Upper extension	Lower extension
$(X_1, X_2, X_3, X_4)^T$	$(X_1, X_2, X_3)^T$	$(X_2, X_4)^T$
$(X_1, X_2, X_3)^T$	$(X_2)^T$	$(X_2)^T$
$(X_2, X_4)^T$	$(X_2)^T$	empty
(X_2)	empty	empty

Fig. 7.3. The extension hierarchy created by the example portfolio $\mathbf{X}$ of Fig. 7.2

The upper and lower extensions $\mathbf{B}_U(\mathbf{X})$ and $\mathbf{B}_D(\mathbf{X})$ of the full portfolio contain themselves barrier options. Thus, additional lattice instances covering the extensions of $\mathbf{B}_U(\mathbf{X})$ and $\mathbf{B}_D(\mathbf{X})$ need to be created as well. Recursion leads to the four partial portfolios shown in Figure 7.3. Altogether, four lattice instances are needed to solve the worst-case pricing problem for the example portfolio.

Definition 8 (Extension hierarchy). *Let* $\mathbf{X}$ *be a portfolio with* $k > 0$ *instruments. Let* $\mathcal{B}$ *denote the set of all partial portfolios of* $\mathbf{X}$. *The extension hierarchy of* $\mathbf{X}$, *written* $\mathcal{B}(\mathbf{X})$, *is defined as the smallest subset of* $\mathcal{B}$ *such that*

- $\mathbf{X} \in \mathcal{B}(\mathbf{X})$, *and*

Input: portfolio **X**
Output: extension hierarchy $\mathcal{B}(\mathbf{X})$

1. Set $\mathcal{B}(\mathbf{X}) := \{\mathbf{X}\}$
2. Repeat the following:
   a) Set $\mathcal{B}' := \bigcup_{\mathbf{Y} \in \mathcal{B}(\mathbf{X})} \{\mathbf{B}_D(\mathbf{Y}), \mathbf{B}_U(\mathbf{Y})\}$
   b) Set $\mathcal{B}' := \mathcal{B}' \setminus (\mathcal{B}(\mathbf{X}) \cup \{\emptyset\})$
   c) Set $\mathcal{B}(\mathbf{X}) := \mathcal{B}(\mathbf{X}) \cup \mathcal{B}'$
   until $\mathcal{B}' = \emptyset$

Fig. 7.4. Finding the extension hierarchy $\mathcal{B}(\mathbf{X})$ amounts to computing a closure. In step 2a, we make sure we know all extensions immediately reachable from the current configuration. In step 2b, extensions that are already known are discarded, as well as the empty extension for which no lattice instance is created

- $\mathbf{Y} \in \mathcal{B}(\mathbf{X})$ implies $\mathbf{B}_U(\mathbf{Y}) \in \mathcal{B}(\mathbf{X})$ and $\mathbf{B}_D(\mathbf{Y}) \in \mathcal{B}(\mathbf{X})$, assuming those are nonempty

Figure 7.4 sketches the algorithm to find the extension hierarchy $\mathcal{B}(\mathbf{X})$ of any given portfolio **X** on a very high level. The sketch is inefficient, but finding $\mathcal{B}(\mathbf{X})$ is the least costly operation in solving the worst-case for **X**. (In our actual implementation, we do employ a more efficient procedure.)

Once $\mathcal{B}(\mathbf{X})$ is known, lattice instances can be created, with appropriately instantiated signatures.

$\mathcal{B}(\mathbf{X})$ is exhaustive. No more lattice instances are required to solve the worst-case pricing problem for **X**. The solution itself is obtained by solving worst-case pricing problems on all lattice instances, transferring boundary data where necessary. The policy outlined in Sect. 5.1 to ensure external consistency leads to the approach shown in Fig. 7.5, outlined on a very high level. (In a concrete implementation, step 3c is done time slice by time slice; the inner loop implicit in step 3c and the outer loop in step 3 change places.)

7.1.2 Complexity

The example in Fig. 7.2 requires four lattice instances for the solution of the worst-case problem. Now consider a second portfolio $\mathbf{X}'$ also consisting of 4 double-barrier options, with the barriers rearranged as shown in Fig. 7.6. In this case, the application of the algorithm in Fig. 7.4 yields an extension hierarchy of 10 elements, listed in Figure 7.7.

It turns out that $10 = 4 \times (4+1)/2$ is indeed the maximum size of any extension hierarchy for a portfolio with 4 instruments. This result can be generalized to the following proposition, taken from [2].

Proposition 2. *Given a portfolio* **X** *of* $k \geq 1$ *instruments such that*

Input: extension hierarchy $\mathcal{B}(\mathbf{X})$

1. Set $n := |\mathcal{B}(\mathbf{X})|$
2. Find an ordering $\mathbf{Y}_{l_1}, \mathbf{Y}_{l_2}, \ldots, \mathbf{Y}_{l_n}$ of $\mathcal{B}(\mathbf{X})$ such that $|\mathbf{Y}_{l_i}| \leq |\mathbf{Y}_{l_j}|$ for $i < j$
3. Repeat for $i = 1, \ldots, n$:
   a) If $\mathbf{B}_U(\mathbf{Y}_{l_i}) \neq \emptyset$ then access the lattice instance for partial portfolio $\mathbf{B}_U(\mathbf{Y}_{l_i})$, and use it for the boundary condition at $U(\mathbf{Y}_{l_i})$
   b) If $\mathbf{B}_D(\mathbf{Y}_{l_i}) \neq \emptyset$ then access the lattice instance for partial portfolio $\mathbf{B}_D(\mathbf{Y}_{l_i})$, and use it for the boundary condition at $D(\mathbf{Y}_{l_i})$
   c) Solve (4.9) for $(\mathbf{Y}_{l_i}, \lambda_{l_i})$, using the data produced in the previous two steps

Fig. 7.5. Solving the worst-case pricing problem for $\mathbf{X}$ requires solving subordinate worst-case problems in the right order. The particular ordering in step 2 implies that $\mathbf{Y}_{l_n} = \mathbf{X}$

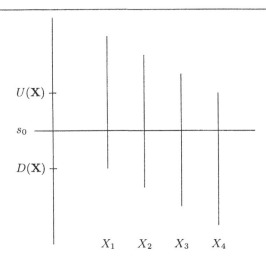

Fig. 7.6. A portfolio $\mathbf{X}' = (X_1, X_2, X_3, X_4)^\mathrm{T}$ consisting of $k = 4$ double-barrier options with up-and-out barriers $u(X_1) > u(X_2) > u(X_3) > u(X_4)$ and down-and-out barriers $d(X_1) > d(X_2) > d(X_3) > d(X_4)$. This particular combination requires 10 lattice instances

$$u(X_1) > u(X_2) > \cdots > u(X_k) \tag{7.13}$$
$$d(X_i) \neq d(X_j) \quad (1 \leq i, j \leq k, i \neq j) \tag{7.14}$$

Then $|\mathcal{B}(\mathbf{X})| \leq k(k+1)/2$.

Proof. By induction over k. For $k = 1$, $|\mathcal{B}(\mathbf{X})| = 1$ by inspection and the proposition is true. Thus assume $k > 1$. Define $\mathbf{X}' = (X_1, \ldots, X_{k-1})^\mathrm{T}$. $\mathbf{X}'$ is

7 Algorithms for Barrier Options

Partial portfolio	Upper extension	Lower extension
$(X_1', X_2', X_3', X_4')^T$	$(X_1', X_2', X_3')^T$	$(X_2', X_3', X_4')^T$
$(X_1', X_2', X_3')^T$	$(X_1', X_2')^T$	$(X_2', X_3')^T$
$(X_2', X_3', X_4')^T$	$(X_2', X_3')^T$	$(X_3', X_4')^T$
$(X_1', X_2')^T$	(X_1')	(X_2')
$(X_2', X_3')^T$	(X_2')	(X_3')
$(X_3', X_4')^T$	(X_3')	(X_4')
(X_1')	empty	empty
(X_2')	empty	empty
(X_3')	empty	empty
(X_4')	empty	empty

Fig. 7.7. The extension hierarchy created by the example portfolio $\mathbf{X}'$ of Fig. 7.6

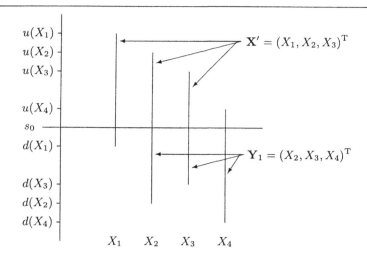

Fig. 7.8. Portfolios $\mathbf{X} = (X_1, X_2, X_3, X_4)^T$ and $\mathbf{X}' = (X_1, X_2, X_3)^T$ illustrate the proof of Prop. 2 for $k = 4$. $\mathbf{Y}_1 = (X_2, X_3, X_4)^T$, the lower extension of $\mathbf{X}$, is also marked. Note that $\mathbf{B}_U(\mathbf{Y}_1) = \mathbf{B}_D(\mathbf{X}') = (X_2, X_3)^T$

a partial portfolio of $\mathbf{X}$ with $k - 1$ instruments and fulfills the premises of the proposition. Figure 7.8 shows an example for $k = 4$.

The idea is to count the lower extensions that must be added to $\mathcal{B}(\mathbf{X}')$ as a consequence of adding X_k to $\mathbf{X}'$. It will turn out that a) all new extensions contain X_k, and b) upper extensions will not cause trouble, thanks to the particular ordering (7.13).

Clearly, $\mathbf{X} \in \mathcal{B}(\mathbf{X})$. By assumption (7.13), $B_U(\mathbf{X}) = \mathbf{X}'$ which implies $\mathbf{X}' \in \mathcal{B}(\mathbf{X})$, and by transitivity $\mathcal{B}(\mathbf{X}') \subset \mathcal{B}(\mathbf{X})$ (here we refer to the algorithm in Fig. 7.4).

Now consider the sequence $\mathbf{Y}_0 = \mathbf{X}$, $\mathbf{Y}_1 = \mathbf{B}_D(\mathbf{Y}_0)$, $\mathbf{Y}_2 = \mathbf{B}_D(\mathbf{Y}_1)$, ..., $\mathbf{Y}_{k-1} = \mathbf{B}_D(\mathbf{Y}_{k-2})$, $\mathbf{Y}_k = \emptyset$. This sequence of consecutive lower extensions has $k+1$ distinct elements, because by assumption (7.14) the d_i's are all distinct.

Thus, $|\mathbf{Y}_i| = k - i$. For $0 \leq i \leq k-1$, define

$$\mathbf{B}_i = \text{select}\,(\mathbf{Y}_i, \{1 \leq j \leq k - i \mid Y_{i,j} \neq X_k\}) \tag{7.15}$$

$\mathbf{B}_i$ is $\mathbf{Y}_i$, possible reduced by the element X_k if it happens to be part of $\mathbf{Y}_i$. Note that by definition of $\mathbf{X}'$,

$$\mathbf{B}_0 = \mathbf{X}' \tag{7.16}$$

We claim that for $0 \leq i \leq k-1$,

$$\mathbf{B}_i \in \mathcal{B}(\mathbf{X}') \tag{7.17}$$

To see this choose $i_0 \in \{0, \ldots, k-1\}$ such that $\mathbf{B}_i \subset \mathbf{Y}_i$ (i.e., X_k is in $\mathbf{Y}_i$, and $\mathbf{B}_i$ is a strictly partial portfolio of $\mathbf{Y}_i$) for $i \leq i_0$ and $\mathbf{B}_i = \mathbf{Y}_i$ (i.e., X_k is not element of $\mathbf{Y}_i$) for $i > i_0$, and note that for $i < i_0$, the equality $\mathbf{B}_D(\mathbf{B}_i) = \mathbf{B}_{i+1}$ holds.

Together with (7.16) as starting point, this implies that $\mathbf{B}_1 \in \mathcal{B}(\mathbf{X}')$ and recursively $\mathbf{B}_i \in \mathcal{B}(\mathbf{X}')$ for $0 \leq i \leq i_0$. Thus, (7.17) is true for at least $0 \leq i \leq i_0$.

That (7.17) is also true for $i_0 + 1 \leq i \leq k - 1$ can be derived from $\mathbf{B}_{i_0} = \mathbf{B}_{i_0+1}$ and $\mathbf{B}_i = \mathbf{Y}_i$ for $i > i_0$, both true by choice of i_0, and since $\mathbf{Y}_i \in \mathcal{B}(\mathbf{X}')$ by definition of $\mathbf{Y}_i$.

$u(X_k)$ is the smallest up-and-out barrier, and X_k is thus the first instrument which is dropped. Therefore, $\mathbf{B}_U(\mathbf{Y}_i) = \mathbf{B}_i$, or $\mathbf{B}_U(\mathbf{Y}_i) \in \mathcal{B}(\mathbf{X}')$ for $0 \leq i \leq k-1$ by (7.17). This implies that the partial portfolios that are not already part of $\mathcal{B}(\mathbf{X}')$ are exactly those that contain X_k, namely $\mathbf{X}, \mathbf{Y}_1, \ldots, \mathbf{Y}_{i_0}$. Thus, since $i_0 \leq k-1$, it follows that the size of $\mathcal{B}(\mathbf{X})$ is bounded by

$$|\mathcal{B}(\mathbf{X})| \leq |\mathcal{B}(\mathbf{X}')| + 1 + (k-1) \tag{7.18}$$

or, by induction,

$$|\mathcal{B}(\mathbf{X})| \leq k(k-1)/2 + 1 + (k-1) = k(k+1)/2 \tag{7.19}$$

which completes the proof. □

It is easy to see that the size of the extension hierarchy does not become larger if $u(X_i)$ and $d(X_i)$ are not distinct as postulated in the proposition, making the upper bound $k(k+1)/2$ the general worst case upper bound for every portfolio of vanilla, single and double barrier options of size k. Motivated by the example in Fig. 7.6, it can be shown that this upper bound is tight.

70   7 Algorithms for Barrier Options

Corollary 1. *Let $\mathbf{X}$ be a portfolio of k double barrier options with barriers $u(X_1) > u(X_2) > \cdots > u(X_k)$ and $d(X_1) > d(X_2) \cdots > d(X_n)$. Then $|\mathcal{B}(\mathbf{X})| = k(k+1)/2$.*

Proof. All elements in the sequence of lower extensions $\mathbf{Y}_0, \ldots, \mathbf{Y}_k$ in the proof of Prop. 2 contain X_k. i.e. $i_0 = k-1$. Since this is true in each inductive step, equality follows in (7.19). □

Most practical cases do not involve double barrier options. Proposition 2 can be specialized for portfolios that contain only single barrier options.

Proposition 3. *Given a portfolio $\mathbf{X}$ of $k \geq 1$ instruments such that*

$$s_0 > d(X_1) > d(X_2) > \cdots > d(X_{k_d}) \qquad (7.20)$$

and

$$u(X_{k_d+1}) > \cdots > u(X_k) > s_0, \qquad (7.21)$$

for some $k_d \in \{0, \ldots, k\}$. Furthermore, $u(X_1), u(X_2), \ldots, u(X_{k_d})$ are very large, and $d(X_{k_d+1}) = \cdots = d(X_k) = 0$. (Thus, there are k_d down-and-out barrier options and $k_u = k - k_d$ up-and-out barrier options in $\mathbf{X}$.) Then

$$|\mathcal{B}(\mathbf{X})| = k_d + k_u + k_d\, k_u. \qquad (7.22)$$

Proof. A simple counting argument will do. If a path $\{S_t(\omega)\}$, $\omega \in \Omega$ of the underlying crosses barrier $d(X_n)$, $1 \leq n \leq k_d$, it must have crossed barriers $d(X_1), \ldots, d(X_{n-1})$ before. Thus, only subsets $X_1, \ldots, X_n$ with contiguous indexes can be knocked out at any particular time. Therefore, we count $k_d + 1$ ways to separate the k_d down-and-out barrier options into knocked-out ones and ones which are still alive.

Similarly, we count $k_u + 1$ ways to divide the k_u up-and-out barrier options in $\mathbf{X}$ into knocked-out and alive ones. Since the up-and-out and down-and-out barrier options are unrelated, there are $(k_d + 1)(k_u + 1)$ combinations altogether. Disregarding the empty combination, we get

$$|\mathcal{B}(\mathbf{X})| = (k_d + 1)(k_u + 1) - 1 = k_d + k_u + k_d\, k_u. \qquad (7.23)$$

□

Proposition 3 shows that the number of lattice instances for a portfolio of single barrier options is linear both in the number of up-and-out resp. down-and-out barrier options. Again, barriers are not necessarily distinct as required in the premise of the proposition. If, however, k_d and k_u are set to the number of distinct up-and-out and down-and-out barriers in $\mathbf{X}$, respectively, then (7.22) remains precise. If $\mathbf{X}$ also contains vanilla options, one additional lattice instance needs to be created, and $|\mathcal{B}(\mathbf{X})| = k_d + k_u + k_d\, k_u + 1$.

7.2 Empirical Results

We apply the theory to example portfolios of barrier options and study the result under two aspects:

– How well do the algorithms in Section (5.2) stand up to practice?
– What are the numerical consequences of introducing uncertainty?

The software Mtg is used to answer these questions.

7.2.1 Numerical Convergence

We first apply Mtg to a double-barrier option and compare the result with two sources in the literature, in order to convince ourselves that the results it delivers are numerically accurate.

Experiment 1: A Double-Barrier Option Set $\mathbf{X} = (X_1)$ and $\lambda_1 = 1$. X_1 is a double barrier option with variable strike K, up-and-out barrier $u = u(X_1)$ and down-and-out barrier $d = d(X_1)$. We assume $S_0 = 2$ and $T = 365$ days. The interest rate r and volatility σ are constant.

For comparison [32] and [58] are used. In [32], a probabilistic approach to price X_1 is suggested. In [58], a pricing formula that consists of a sum of an infinite series is used. Mtg is run with four different time steps $dt_{\max} = 1/(N \times 365)$, where $N = 1, 5, 20$ and 50, respectively, as well as under explicit and Crank-Nicholson schemes (in the explicit scheme $dt = dt_{\max}$ after the algorithm in Fig. 5.4 is run).

Figure 7.9 lists the results for three combinations of r, σ, u and d. Geman and Yor's method in [32] is quoted as "G-Y", and Kunitomo and Ikeda's method in [58] is quoted as "K-I."

The convergence is very satisfactory. For $N = 5$, the first four digits after the decimal point of the results of all three methods agree. The Crank-Nicholson scheme converges slightly faster than the explicit forward Euler scheme. It is, however, between 30 and 50% slower than the explicit scheme.

The theoretical time complexity is $O(N^{3/2})$, due to trimming. Measurements for all N validate the theory and yield a running time of approximately $0.1 \times N^{3/2}$ seconds for the explicit scheme, and $0.14 \times N^{3/2}$ seconds for Crank-Nicholson (on a 166-MHz PC).

Experiment 2: A Portfolio of Single-Barrier Puts To test the algorithm in Fig. 5.5, a portfolio of four down-and-out at-the-money puts is priced, listed in Figure 7.10. All options mature in 30 days. The other parameters are $S_0 = 100$, $r = 0.025$ and $\sigma = 0.2$.

The results for the explicit and the Crank-Nicholson scheme are summarized in Fig. 7.11. Also shown are $\overline{\sigma}_D$ and $\overline{\sigma}_U$ (defined in Sect. 5.2). There are four regions in the lattice with differing $\overline{\sigma}_D$ and $\overline{\sigma}_U$; the three interior barriers at 98, 95 and 90 mark the boundaries between these regions. The

7 Algorithms for Barrier Options

	N (periods per day)	scheme (CR = Crank-Nicholson)	$\sigma = 0.2$ $r = 0.02$ $K = 2$ $d = 1.5, u = 2.5$	$\sigma = 0.5$ $r = 0.05$ $K = 2$ $d = 1.5, u = 3$	$\sigma = 0.5$ $r = 0.05$ $K = 1.75$ $d = 1, u = 3$
Mtg	1	explicit	0.040899	0.017666	0.075914
Mtg	1	CR	0.040968	0.017844	0.076146
Mtg	5	explicit	0.041050	0.017819	0.076102
Mtg	5	CR	0.041063	0.017856	0.076149
Mtg	20	explicit	0.041079	0.017848	0.076158
Mtg	20	CR	0.041083	0.017857	0.07617
Mtg	50	explicit	0.041085	0.017853	0.076168
Mtg	50	CR	0.041086	0.017857	0.076173
G-Y			0.0411	0.0178	0.07615
K-I			0.041089	0.017856	0.076172

Fig. 7.9. Prices obtained for a double barrier call option with each of the three methods Mtg, G-Y and K-I. There is no uncertainty

type	strike	barrier	position
put	100	98	long 200 contracts
put	100	95	long 10 contracts
put	100	90	long 2 conctracts
put	100	85	long 1 contract

Fig. 7.10. A portfolio of four down-and-out 30-day at-the-money puts. The position in each put is approximately inverse proportional to the relative value that it contributes to the portfolio

range of $\overline{\sigma}_D$ and $\overline{\sigma}_U$ narrows as N becomes large, from a range of 0.29–0.38 with an absolute difference of 0.09 for $N = 1$ to a range of 0.28285–0.028591 with an absolute difference of only 0.000306 for $N = 400$. (It is obvious that smaller and thus more numerous spatial increments have to be bent relatively less to match the barriers.)

A closed form formula for down-and-out barrier puts yields 10.287 as the model value. The numerical result is sufficiently close for $N \geq 100$.

Here we measure a running time of $0.006 \times N^{3/2}$ (explicit) respectively $0.0162 \times N^{3/2}$ seconds (Crank-Nicholson). Crank-Nicholson trails the explicit scheme by a factor of ≈ 2.7, while not yielding significant higher accuracy.

Note that this example does not exhibit uncertainty. Only one lattice instance is needed to compute the price of the portfolio, and in this case the individual values of the puts might as well have been computed separately and added up. We add uncertainty to this particular portfolio below.

N (periods per day)	price explicit	price Crank-Nicholson	$\bar{\sigma}_D, \bar{\sigma}_U$ between 98 and above	$\bar{\sigma}_D, \bar{\sigma}_U$ between 98 and 95	$\bar{\sigma}_D, \bar{\sigma}_U$ between 95 and 90	$\bar{\sigma}_D, \bar{\sigma}_U$ between 90 and below
1	7.72429	7.76010	0.38597	0.29699	0.34432	0.36400
2	7.79286	7.80849	0.33426	0.34294	0.29819	0.31524
5	9.96666	10.0005	0.28769	0.33205	0.28872	0.30523
10	10.1146	10.1320	0.30514	0.31306	0.29695	0.28777
20	10.2094	10.2185	0.28769	0.29515	0.28872	0.28727
50	10.2532	10.2569	0.30325	0.30001	0.29216	0.28599
100	10.2709	10.2727	0.29690	0.28285	0.28693	0.28737
200	10.2793	10.2803	0.28729	0.28966	0.28643	0.28599
400	10.2832	10.2837	0.28591	0.28285	0.28300	0.28364

Fig. 7.11. Results for a portfolio of four down-and-out at-the-money puts. The model value is 10.287. Also shown are $\bar{\sigma}_D$ and $\bar{\sigma}_U$ for the four significant regions of the lattice, determined by the interior barriers 98, 95 and 90 and found by the algorithm in Fig. 5.5

type	strike	position	U&O barrier	D&O barrier
call	110	long 1 contract	120	90
put	100	long 1 contract	–	95

type	strike	position	quoted price
call	110	long 0.24 contracts	17% implied vol
call	100	short 0.98 contracts	13% implied vol
call	90	long 8.47 contracts	15% implied vol

Fig. 7.12. The portfolio consists of two 30-day barrier options and four 30-day vanilla options. The market prices for the vanilla options are quoted as implied volatility. The contribution of X_3, X_4 and X_5, given their market prices and positions, is 84.499

7.2.2 Introducing Uncertainty

Experiment 3: Two Barrier Options Hedged We introduce uncertainty in σ for the first time and choose $\sigma_{\min} = 0.1$ and $\sigma_{\max} = 0.2$ as upper and lower bounds. The other parameters are $S_0 = 100$ and $r = 0.02$.

Let **X** be a portfolio of 5 instruments. X_1 is a double-barrier call, X_2 is a single-barrier put, and X_3, X_4 and X_5 are vanilla calls whose market prices are known. All options mature in 30 days. Let $\lambda = (1, 1, 0.24, -0.98, 8.47)^T$ be the position in **X**.

Four lattice instances are necessary to solve the worst-case problem for $(\mathbf{X}, \lambda)$. Their signatures are, respectively,

N (periods per day)	price explicit	Crank-Nicholson	explicit price minus premium for X_3, X_4, X_5
5	85.5709	85.5720	1.0719
10	85.5762	85.5771	1.0772
20	85.5788	85.5791	1.0798
50	85.5801	85.5803	1.0811

Fig. 7.13. Worst-case prices for a double-barrier option, a single-barrier option and three traded vanillas. The last column shows the contribution of X_1 and X_2 to the worst-case price, given that the market prices for X_3, X_4 and X_5 are 17, 13 and 15% implied volatility, respectively

- the two barrier options plus the vanillas;
- the double barrier option plus the vanillas;
- the single barrier option plus the vanillas;
- the vanillas.

The worst-case price of $(\mathbf{X}, \lambda)$ is shown in Fig. 7.13 for various time steps. Results under the explicit scheme and Crank-Nicholson are in close agreement, although Crank-Nicholson turns out to be between 2 and 4 times slower.

Given their market prices, the transaction involving X_3, X_4 and X_5 creates a premium of 84.499. Thus, anyone charging at least 85.5801 for the entire package and at the same time entering the offsetting position in X_3, X_4 and X_5 (thus effectively charging 1.0811 for X_1 and X_2) will break even or make a profit provided the volatility stays within the band $0.1 \leq \sigma \leq 0.2$ over the next 30 days. It can be shown that our particular offsetting position in the vanillas is optimal in the sense that 1.0811 is the smallest surcharge for X_1 and X_2 for any offsetting position. The position $(0.24, -0.98, 8.47)$ in (X_3, X_4, X_5) is the optimal hedge portfolio for the position $(1, 1)$ in (X_1, X_2). See also Sect. 4.2.2.

Experiment 4: Single-Barrier Portfolios of Various Sizes According to Prop. 3, the running time complexity is $O(k_d + k_u + k_d k_u)$ in the number of down-and-out barriers k_d and the number of up-and-out barriers k_u, assuming there are no double-barrier options in the portfolio. We want to validate this formula experimentally.

We augment the portfolio of four down-and-out barrier puts in Fig. 7.10 by four up-and-out barrier calls as shown in Fig. 7.14.

For each combination

$$(k_d, k_u) \in \{(x, y) \mid 0 \leq x, y \leq 4, x \geq 1, y \leq x\} \tag{7.24}$$

of down-and-out and up-and-out barrier options, the worst-case price is computed for the portfolio consisting of the first k_d puts and the first k_u calls.

7.2 Empirical Results   75

type	strike	barrier	position
put	100	98	long 200 contracts
put	100	95	long 10 contracts
put	100	90	long 2 conctracts
put	100	85	long 1 contract
call	100	102	long 200 contracts
call	100	105	long 10 contracts
call	100	110	long 2 conctracts
call	100	115	long 1 contract

Fig. 7.14. The portfolio of four down-and-out 30-day at-the-money puts of Fig. 7.10, augmented by four up-and-out 30-day at-the-money calls

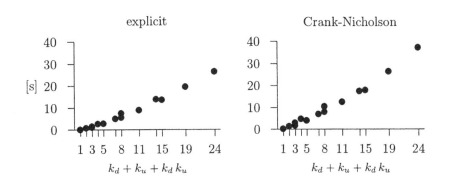

Fig. 7.15. The running times in seconds under the explicit resp. Crank-Nicholson schemes, plotted against the number of lattice instances necessary to process various combinations of single-barrier options

Overall, 14 option combinations are priced. Since we are only interested in the running time, we set $\sigma_{\min} = 0.199999$ and $\sigma_{\max} = 0.2$ just to make sure the problem becomes nonlinear for the pricing software. The other parameters are $S_0 = 100$, $r = 0.025$ and $N = 20$ (i.e., $dt = 1/(20 \times 365)$).

Figure 7.15 plots the running times against the number of lattice instances $k_d + k_u + k_d k_u$ required for the respective option combination. The graphic shows that the running time progresses linearly, thereby validating Prop. 3 experimentally. (Slight deviations are expected: each partial portfolio has its unique boundary, and the corresponding lattice instance thus a unique continuation region. Numerical processing concentrates on the continuation region, causing slightly different processing times for different lattice instances.)

8 Algorithms for American Options

Similar to barrier options, American options may be terminated prematurely sometime between the settlement and the face maturity date. This seems to make the techniques discussed in Chapters 5 and 7 applicable to portfolios that contain American options as well.

This is true—in principle. The fundamental difference is that the early exercise of American options is voluntary. The precise date is not known a priori, although it may be assumed that holders of American options time early exercise so as to maximize their expected payoff. How mathematical finance models this behavior can be read in [11] and [54]. Numerically, American options can be evaluated with projected SOR (Successive Over-Relaxation) methods on a lattice or tree, as described in [78]. Other approaches are also possible (see [62], for instance).

Uncertainty in some of the model coefficients adds another twist to the problem: the early exercise strategy for an individual American option X_1 now depends on the entire position $(\mathbf{X}, \lambda)$, not merely on the contribution of X_1 judged separately. We have encountered this situation with barrier options: once X_1 is exercised, the exposure of the remaining partial portfolio to fluctuations in the uncertain coefficients may be different, and so may the worst-case value. Thus it is not possible to pre-process the components of $\mathbf{X}$ with American early exercise features separately and use the so found early exercise boundaries just like knock-out boundaries are used in the case of barrier options. Rather, the continuation and early exercise regions for each American option in $\mathbf{X}$ must be searched for dynamically, by considering the consequences of all possible early exercise strategies (of which there are plenty if $\mathbf{X}$ contains several American options) on the worst-case value of the entire position.

In this chapter, we show how to implement the dynamic search for continuation and early exercise regions within the framework of worst-case pricing. We also show how to cope with the explosion of combinations: it is possible to reduce the combinatorial complexity in most practical cases considerably.

In some sense, the concept of optional early exercise *is* merely an extension of the notion of forced knock-out. The algorithms in this chapter work for both American and barrier options indiscriminately. In particular, they are capable of pricing a portfolio of American barrier options.

Once again, σ shall be the only uncertain model coefficient.

8.1 Early Exercise Combinations

We use the lattice approach of Chapter 5. Let $(\mathbf{X}, \lambda)$ be the portfolio, and assume all k instruments in $\mathbf{X}$ mature at time T, and may be exercised early at any time between now and T.

The local data flow shown in Figs. 5.2 and 5.3 captures the situation only partially if American options are present. $\hat{V}(j, i; L)$, unmodified, represents the worst-case portfolio value under the restriction that no option be exercised at $t = t_i$ and $S_i = s_j$ (recall that S_i is an abbreviation for S_{t_i} and s_j is the jth spatial level of the lattice). $\hat{V}(j, i; L)$ needs to be compared to other worst-case values that arise under viable early exercise combinations, and a proper selection needs to be made. Thus, the original scheme turns into a two-tiered numerical-combinatorial regime:

1. The finite-difference scheme is applied to find the worst-case value under a no-exercise assumption. This is the numerical phase.
2. This preliminary value competes against the worst-case values delivered by all viable early exercise combinations. It may or may not be updated. This is the combinatorial phase.

$\hat{V}(j, i; L)$ is always paired with its gradient vector $(\hat{v}_1(j, i; L), \ldots, \hat{v}_k(j, i; L))^{\mathrm{T}}$. Although sometimes not explicitly mentioned, the gradient is always computed, and modified, together with $\hat{V}(j, i; L)$.

Figure 8.1 presents this approach graphically. Subordinate lattice instances need to be accessed in the combinatorial phase. As some early exercise combinations might be dismissed right away, exactly which lattice instances must be available at a given node instance $(j, i; L)$ is determined at runtime. Clever selection techniques lead to significant speedup.

Note that Fig. 8.1 is correct for explicit finite difference schemes, but not necessarily for mixed explicit/implicit schemes such as Crank-Nicholson. As the update of one $\hat{V}(j, i; L)$ affects all the others implicitly, iterative improvement over both phases 1 and 2, applied to all node instances $(\cdot, i; L)$ of the current time slice, is required. A modified projective SOR method, for instance, may do.

For this reason, all experimental results were obtained with explicit forward Euler. Although we have implemented Crank-Nicholson (and projected SOR), we focus on combinatorial aspects in this chapter and ignore the numerical side as much as possible.

8.1.1 Long and Short Positions

Which early exercise combinations should be adopted at (j, i, L)? Which combination is "suitable", in the language of Fig. 8.1? Simply choosing the one

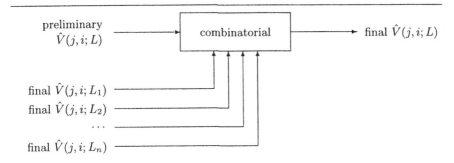

Fig. 8.1. The preliminary worst-case value $\hat{V}(j,i;L)$ at node instance $(j,i;L)$, result of the dataflow in Fig. 5.2, enters the combinatorial post-processor which selects a suitable early exercise combination. To do that it needs to access lattice instances $L_1, \ldots, L_n$, all carrying partial portfolios of $\mathbf{X}$

that represents the largest worst-case value is not sufficient, since control lies not only with the other party to whom some of the American instruments in the portfolio have been sold, but also with the agent who may *own* some of the American options.

The Worst-Case Price Revisited To clarify this point, we remind ourselves that

> the worst-case price of $(\mathbf{X}, \lambda)$ represents the largest amount of funds that may be necessary to delta-hedge a portfolio $(\mathbf{X}, \lambda)$. It thus represents the safe price which, when charged, guarantees that the seller of $(\mathbf{X}, \lambda)$ will not incur any losses.

The worst-case price therefore represents the point of view of the sell-side:

- $\lambda_i > 0$ means that the agent has sold λ_i contracts in the ith instrument and does therefore not control the early exercise strategy for this instrument. (When the sell-side sells a long position in X_i, it effectively goes short X_i.)
- $\lambda_i < 0$ means that the agent has bought $|\lambda_i|$ contracts in the ith instrument and therefore controls its early exercise strategy.

The worst-case value of $(\mathbf{X}, \lambda)$ is the worst-case liability of the sell-side. Positive values mean that additional funds must be provided to hedge against the worst-case. They represent an upper bound for the price (i.e., the most desirable price) the seller can justifiably charge the buyer, assuming the buyer agrees with the seller on the uncertain model coefficients.

Similarly, negative values indicate a net flow of funds from the seller to the buyer; the absolute value represents a lower bound (corresponding to the unaltered value being an upper bound) on the amount of funds p the seller needs to transfer together with $(\mathbf{X}, \lambda)$. Any amount smaller than p is

80   8 Algorithms for American Options

no longer competitive under the particular uncertainty assumptions of the model. Thus, p is the most desirable cost for the seller.

The Best Worst-Case Price In the previous paragraph we gave an economic justification for the maximum-principle in worst-case pricing. So far, however, agents from whose perspective the worst-case price is computed have been unable to modify their risk profile after the position $(\mathbf{X}, \lambda)$ has been set up.

This is different if $\mathbf{X}$ contains American instruments. Suppose X_i is American, and $\lambda_i < 0$. Define $\mathbf{X}' = (X_1, \ldots, X_{i-1}, X_{i+1}, \ldots, X_k)^{\mathrm{T}}$, and λ' accordingly. Define $\mathbf{X}^e = \mathbf{X}$, with X_1 modified to preclude early exercise if $t = t_i$ (early exercise is still allowed for $t \geq t_{i+1}$). Let the signature of lattice instance L^e and L' be $(\mathbf{X}^e, \lambda)$ and $(\mathbf{X}', \lambda')$, respectively.

Consider the two quantities

$$V_1 = \hat{V}(j, i; L^e)$$
$$V_2 = \hat{V}(j, i; L') + \lambda_i X_i \qquad (8.1)$$

Both are worst-case prices: V_1 under the restriction that X_1 must not be exercised now, but may be later, and V_2 under the constraint that X_1 be forcibly exercised now. Let us assume that there are no other American options in $\mathbf{X}$. Then V_1 and V_2 exhaust the possible early exercise combinations for X_1 and the remainder of the portfolio, and we may express $\hat{V}(j, i; L)$ in terms of V_1 and V_2.

Since the agent may *choose* between V_1 and V_2, it is advisable to choose the *minimum*. This strategy reduces the worst-case liability of the agent and assures that the agent selects the most competitive price when $(\mathbf{X}, \lambda)$ is offered for sale, while still being hedged against volatility fluctuations.

Definition 9 (Best worst-case). *The best worst-case value of portfolio $(\mathbf{X}, \lambda)$ is the minimal worst-case value of $(\mathbf{X}, \lambda)$, where the minimum is taken over the early exercise strategies open to the seller of $(\mathbf{X}, \lambda)$.*

We illustrate the principle of best worst-case evaluation with an example, and give a formalization in Sect. 8.1.2.

Assume $\mathbf{X} = (X_1, X_2)^{\mathrm{T}}$ and $\lambda = (-1, 1)^{\mathrm{T}}$. Both instruments are American. The seller of $(\mathbf{X}, \lambda)$ controls early exercise for X_1, but is subjected to any early exercise decisions made by the holder of X_2. Suppose the outcomes of the four early exercise combinations at node instance (j, i) are those shown in Fig 8.2:

- 40 if X_1 and X_2 are both exercised (payoff of X_1 plus payoff of X_2);
- 10 if only X_1 is exercised (payoff of X_1, plus worst-case value of X_2 under the restriction that X_2 not be exercised at node (j, i));
- 20 if only X_2 is exercised (payoff of X_2, plus best worst-case value of X_1 under the restriction that X_1 not be exercised at node (j, i));

	exercise X_2	don't exercise X_2
exercise X_1	40	10
don't exercise X_1	20	$\boxed{30}$

Fig. 8.2. Early exercise combinations at node (j,i) and their corresponding values. Bold values are row maxima; the framed value is the best worst-case

---

- 30 if neither instrument is exercised (best worst-case value under the restriction that X_1 and X_2 not be exercised at node (j,i)).

Definition 9 requires a strategy for X_1 that guarantees the lowest value under all possible decisions of the holder of X_2. The agent therefore selects the strategy with the lowest row maximum. In the example, the agent postpones the exercise of X_1 at least until time t_{i+1} and thus guarantees a worst-case value of 30.

8.1.2 Best Worst-Case Evaluation Formalized

Most of the exposition in this section is taken from [17] and reformulated for a discrete setting.

Some notational remarks. In general, if nothing else is said, the lattice instance associated with signature $(\mathbf{X}, \lambda)$ is denoted by L, and vice versa. Its size is denoted $|L| = |\mathbf{X}| = |\lambda| = k$. In some cases, however, $\mathbf{X}_L$ and λ_L express this relationship explicitly. If $(\mathbf{X}', \lambda') \subseteq (\mathbf{X}, \lambda)$ is partial, and L' is the corresponding lattice instance, we refer to L' as a sub-lattice instance. L is sometimes called the root-lattice instance.

Definition 10 (Separation into long and short). *Let $(\mathbf{X}', \lambda') \subseteq (\mathbf{X}, \lambda)$ be a partial portfolio of size k'. Then*

$$\text{long}\,(\mathbf{X}', \lambda') = \{1 \leq n \leq k' \mid \lambda'_n < 0 \text{ and } X'_n \text{ is American}\}$$
$$\text{short}\,(\mathbf{X}', \lambda') = \{1 \leq n \leq k' \mid \lambda'_n > 0 \text{ and } X'_n \text{ is American}\} \quad (8.2)$$
$$\text{am}\,(\mathbf{X}') = \{1 \leq n \leq k' \mid \lambda'_n \neq 0 \text{ and } X'_n \text{ is American}\}$$

separate the American instruments in $\mathbf{X}'$ into long and short positions. (Recall that for the sell-side, $\lambda'_n > 0$ translates to X'_n being sold.)

Definition 11 ("Europeanization"). *Let $\mathbf{X}$ be a portfolio, and $X_n \in \text{am}\,(\mathbf{X})$ one of its American instruments. Then X_n^E is its "europeanized" version: early exercised is precluded everywhere. If G is any process involving X_n, then we write G^E for the corresponding process involving X_n^E (where "corresponding" depends on the context). Similarly, L^E is a lattice instance whose signature contains europeanized versions $X_1^E, \ldots$ of American instruments $X_1, \ldots$.*

8 Algorithms for American Options

Definition 12 (Residual lattice instance). *Let L be a lattice instance with signature $(\mathbf{X}, \lambda)$, and let $M \subseteq \{1, \ldots, |L|\}$. Then $\neg M = \{1, \ldots, |L|\} \setminus M$, and $L_{\neg M}$ with signature*

$$(\text{select}(\mathbf{X}, \neg M), \text{select}(\lambda, \neg M)) \tag{8.3}$$

is called the residual lattice instance *of L.*

Definition 13 (Payoff from early exercise). *Given a lattice instance L with signature $(\mathbf{X}, \lambda)$ and an enumeration of instruments $M \subseteq \text{am}(\mathbf{X})$. Then the payoff from early exercise of the instruments in M is given by the linear combination*

$$\text{payoff}(L, M) = \text{select}(\lambda, M) \cdot \text{select}(\mathbf{X}, M) \tag{8.4}$$

The Best Worst-Case Price Process

Definition 14 (Local fixation of early exercise). *Let $\hat{F} = \{\hat{F}_i\}$ be any discrete process defined on the space of node instances. $\hat{F}_i(L)$ is a random variable; $\hat{F}_i(j; L)$ its value at node instance $(j, i; L)$. Assume $(\mathbf{X}, \lambda)$ is the signature of L, and choose $\sigma \in \mathcal{C}$.*

Then we define the local fixation *of $\hat{F}$ for $M \subseteq \text{am}(\mathbf{X})$ as follows:*

$$F_i(L, M, \sigma) = \frac{1}{\beta_i} \mathbb{E}_{Q(\sigma)} \left(\beta_{i+1} \hat{F}_{i+1}(L_{\neg M}) \mid \mathcal{F}_i \right) + \text{payoff}(L, M) \tag{8.5}$$

where $L_{\neg M}$ is the residual lattice instance of L. $F_i(L, M, \sigma)$ is also a random variable, and $F_i(j; L, M, \sigma)$ its value at node instance $(j, i; L)$.

In some sense, the local fixation $F_i(L, M, \sigma)$ "harnesses" the power of $\hat{F}_i(L)$ by fixing the volatility σ as well as the early exercise strategy for one time period. The parameter M in Def. 14 has the effect of modifying the features of $\mathbf{X}$ locally. The maturity date of the instruments covered by M is advanced to t_i, and for all other instruments the earliest date on which early exercise is permissible is set to t_{i+1}. It is easy to see that F is adaptable.

The following definitions remove the restrictions on the parameters M and σ in $F(L, M, \sigma)$ again. The result will not be the original $\hat{F}$, but a version that incorporates the best worst-case paradigm in Def. 9.

Definition 15 (Local uncertainty reintroduced). *Let $F(L, M, \sigma)$ be a local fixation. Then $\hat{F}(L, M) = \{\hat{F}_i(L, M)\}$, defined as*

$$\hat{F}_i(L, M) = \sup_{\sigma \in \mathcal{C}} F_i(L, M, \sigma) \tag{8.6}$$

reintroduces uncertainty in σ locally.

8.1 Early Exercise Combinations 83

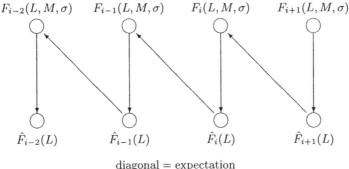

diagonal = expectation
vertical = selection

Fig. 8.3. An illustration of Def. 14. The process $F(L, M, \sigma) = \{F_i(L, M, \sigma)\}$ depends on the values the base process $\hat{F}(L) = \{\hat{F}_i(L)\}$ takes on in subsequent time slices, by taking the (discounted) expectation over all possible transitions. Although not implied in Def. 14, we shall see later that $\hat{F}(L)$ in turn may depend on $F(L, M, \sigma)$, by selecting among feasable instantiations of M and σ

Definition 16 (Local optionality reintroduced). *Let $\hat{F} = \{\hat{F}_i\}$ be any discrete process defined on the space of node instances. Let $\hat{F}(L, M, \sigma)$ be its local fixation, and let $\hat{F}(L, M)$ be the process defined in Def. 15. Let*

$$A(L) = \text{long}\,(\mathbf{X}, \lambda)$$
$$B(L) = \text{short}\,(\mathbf{X}, \lambda) \tag{8.7}$$

denote the American instruments on lattice instance L with signature $(\mathbf{X}, \lambda)$. Then the process $\hat{G}(L) = \{\hat{G}_i(L)\}$, defined as

$$\hat{G}_i(L) = \min_{A \subseteq A(L)} \max_{B \subseteq B(L)} \hat{F}_i(L, A \cup B) \tag{8.8}$$

is said to reintroduce optionality *locally.*

Definition 17 (Best worst-case process). *Given $(\mathbf{X}, \lambda)$. Let $\hat{F} = \{\hat{F}_i\}$ be a discrete process defined on the space of node instances belonging to the set of root- and sub-lattice instances for $(\mathbf{X}, \lambda)$, and let $\hat{G}_i$ reintroduce optionality locally. Then $\hat{F}$ is called a* best worst-case price process *process if*

$$\hat{F}_i(L) = \hat{G}_i(L) \tag{8.9}$$

for all lattice instances L with signatures $(\mathbf{X}', \lambda')$, $\mathbf{X}' \subseteq \mathbf{X}$ and $\lambda' \in \mathbb{R}^{|\mathbf{X}'|}$, and $0 \leq i \leq N - 1$. If furthermore the payoff condition

$$\hat{F}_N(L) = \lambda_L \cdot \mathbf{X}_L \tag{8.10}$$

holds for all those L, then $\hat{F}$ is called a best worst-case price *process for $(\mathbf{X}, \lambda)$.*

84   8 Algorithms for American Options

We require (8.9) and the payoff condition to hold for all possible positions, not only for those which can be constructed by removing elements from λ. This is necessary because auxiliary positions (especially in singleton partial portfolios) may be required to support the computation.

The vertical arrows in Fig. 8.3 now make sense: the process $\hat{F}$ in the picture is a best worst-case price process.

Notice how Defs. 15 and 16 implement the proper hierarchy of local minimization and maximization, corresponding to the following sequence of moves between the agent, the client(s) who hold the instruments sold by the agent, and the market:

1. The agent chooses instruments to exercise. Candidates are found in long $(\mathbf{X}, \lambda)$.
2. Similarly, the clients decide for instruments enumerated in short $(\mathbf{X}, \lambda)$.
3. The market acts by "selecting" the volatility of the underlying asset until the subsequent time slice.

No other order is plausible. The minimization operator in (8.8) guarantees that the agent makes the best possible choice, assuming maximal adversion by the client(s) and by the market.

Proposition 4 (Uniqueness). *There is only one best worst-case price process for $(\mathbf{X}, \lambda)$.*

Proof. Let $\hat{F}$ and $\hat{G}$ be both best worst-case price processes for $(\mathbf{X}, \lambda)$. Since, by definition, both processes fulfill the payoff condition (8.10), they agree for $i = N$. Induction over $i = N - 1, \ldots, 0$ and the definiteness of Defs. 15 and 16, including (8.8), imply uniqueness. $\square$

Compatibility with Traditional Formulations It is easy to verify that the definition of the best worst-case price process $\hat{F}$ for $(\mathbf{X}, \lambda)$ implies that

$$\hat{F}_i(L) = \sup_{\sigma \in \mathcal{C}} \frac{1}{\beta_i} \mathrm{E}_{Q(\sigma)} \left(\beta_{i+1} \hat{F}_{i+1}(L) \mid \mathcal{F}_i \right) \quad (8.11)$$

if long $(\mathbf{X}, \lambda)$ = short $(\mathbf{X}, \lambda) = \emptyset$, i.e. if all instruments in $\mathbf{X} = \mathbf{X}^E$ are European. Expansion leads to

$$\begin{aligned}\hat{F}_i(L) &= \sup_{\sigma \in \mathcal{C}} \frac{1}{\beta_i} \mathrm{E}_{Q(\sigma)} \left(\sup_{\sigma' \in \mathcal{C}} \mathrm{E}_{Q(\sigma')} \left(\beta_{i+2} \hat{F}_{i+2}(L) \mid \mathcal{F}_{i+1} \right) \mid \mathcal{F}_i \right) \\ &= \sup_{\sigma \in \mathcal{C}} \frac{1}{\beta_i} \mathrm{E}_{Q(\sigma)} \left(\mathrm{E}_{Q(\sigma')} \left(\beta_{i+2} \hat{F}_{i+2}(L) \mid \mathcal{F}_{i+1} \right) \mid \mathcal{F}_i \right)\end{aligned} \quad (8.12)$$

where we exploit the fact that the outer and inner expectations are taken over distinct periods of time. The outer supremum is insensitive to changes of the values of the function σ for $t \notin [t_i, t_{i+1})$, and the inner supremum is insensitive to changes of the values of σ' for $t \notin [t_{i+1}, t_{i+2})$. Both operators can thus be merged. Telescoping leads to

$$\hat{F}_i(L) = \sup_{\sigma \in \mathcal{C}} \frac{1}{\beta_i} \mathrm{E}_{Q(\sigma)} \left(\beta_{i+2} \, \hat{F}_{i+2}(L) \mid \mathcal{F}_i \right)$$

$$\cdots$$

$$= \sup_{\sigma \in \mathcal{C}} \frac{1}{\beta_i} \mathrm{E}_{Q(\sigma)} \left(\beta_N \, \hat{F}_N(L) \mid \mathcal{F}_i \right) \quad (8.13)$$

$$= \sup_{\sigma \in \mathcal{C}} \frac{1}{\beta_i} \mathrm{E}_{Q(\sigma)} \left(\beta_N \left(\lambda \cdot \mathbf{X} \right) \mid \mathcal{F}_i \right)$$

This is the discrete version of (4.10) in Fact 5.

Now assume long $(\mathbf{X}, \lambda) = \emptyset$, but short $(\mathbf{X}, \lambda) \neq \emptyset$, i.e. some of the options sold are American, but none of the options held long (by the sell-side) are. In this case, the minimization operator in (8.8) is superfluous, and we get, after unrolling the definitions,

$$\hat{F}_i(L) = \max_{B \subseteq B(L)} \sup_{\sigma \in \mathcal{C}} \frac{1}{\beta_i} \mathrm{E}_{Q(\sigma)} \left(\beta_{i+1} \hat{F}_{i+1}(L_{\neg B}) \mid \mathcal{F}_i \right) + \mathrm{payoff}\,(L, B) \quad (8.14)$$

There are two cases:

1. The maximum is attained at $B = \emptyset$, or $\hat{F}_i(L) = \hat{F}_i(L, \emptyset)$. In this case, early exercise of any American instrument does not make the situation worse, and we conclude

$$\hat{F}_i(L) = \sup_{\sigma \in \mathcal{C}} \frac{1}{\beta_i} \mathrm{E}_{Q(\sigma)} \left(\beta_{i+1} \hat{F}_{i+1}(L) \mid \mathcal{F}_i \right)$$
$$= \sup_{\sigma \in \mathcal{C}} \frac{1}{\beta_i} \mathrm{E}_{Q(\sigma)} \left(\beta_{i+1} \max_{B \subseteq B(L)} \hat{F}_{i+1}(L, B) \mid \mathcal{F}_i \right) \quad (8.15)$$

If, in turn, $\hat{F}_{i+1}(L) = \hat{F}_{i+1}(L, \emptyset)$ for all s_j, conditioned on $\mathcal{F}_i$, then

$$\hat{F}_i(L) = \sup_{\sigma \in \mathcal{C}} \frac{1}{\beta_i} \mathrm{E}_{Q(\sigma)} \left(\beta_{i+2} \max_{B \subseteq B(L)} \hat{F}_{i+2}(L, B) \mid \mathcal{F}_i \right) \quad (8.16)$$

and so on. Most of the time this will not be so, however, and branching into case 2 may occur, depending on $\mathcal{F}_{i+1}$.

2. The maximum is attained for some $B \neq \emptyset$, or $\hat{F}(L) \neq \hat{F}(L, \emptyset)$. Let $L' = L_{\neg B}$ be the residual lattice instance, with signature $(\mathbf{X}', \lambda') = (\mathbf{X}_B, \lambda_B)$. Now choose $B' \subseteq \mathrm{short}\,(\mathbf{X}', \lambda')$ such that a)

$$\hat{F}_i(L') = \sup_{\sigma \in \mathcal{C}} \frac{1}{\beta_i} \mathrm{E}_{Q(\sigma)} \left(\beta_{i+1} \hat{F}_{i+1}(L'_{\neg B'}) \mid \mathcal{F}_i \right) + \mathrm{payoff}\,(L', B') \quad (8.17)$$

and b) no subset B'' for which (8.17) also holds has fewer elements. Then $B' = \emptyset$ must necessarily be true, for otherwise one could identify the instruments enumerated in B' in the original portfolio $\mathbf{X}$ (the indexing might be different). They wouldn't be covered by B since

$B' \cap B = \emptyset$, and exercising them in addition to those in B would increase the worst-case value, contradicting the maximality under B in (8.14). (The argument uses the associativity of the maximum operator.) Thus, $\hat{F}_i(L') = \hat{F}_i(L', \emptyset)$, and

$$\hat{F}_i(L) = \hat{F}_i(L', \emptyset) + \text{payoff}\,(L, B) \tag{8.18}$$

We have found the *free boundary*.

These two cases can be combined with the introduction of a discrete stopping time τ_i such that

$$\tau_i(L) = \inf\left\{i \le u \le N \mid \hat{F}_u(L) \ne \hat{F}_u(L, \emptyset) \text{ or } u = N\right\} \tag{8.19}$$

$\tau_i(L)$ marks the first time case 2 is encountered on lattice instance L. Combining cases 1 and 2, (8.14) can be re-written

$$\hat{F}_i(L) = \sup_{\sigma \in \mathcal{C}} \frac{1}{\beta_i} \mathrm{E}_{Q(\sigma)}\left(\beta_{\tau_i(L)} \max_{B \subseteq B(L)} \left[\hat{F}_{\tau_i(L)}(L_{\neg B}) + \text{payoff}\,(L, B)\right] \,\bigg|\, \mathcal{F}_i\right) \tag{8.20}$$

which is the form for the early exercise problem that can be found in standard textbooks, such as [26] (in a linear setting).

Sub-Additivity Revalidated Fact 6 states that worst-case prices are sub-additive. In the following paragraphs, we show that best worst-case prices have the same property.

Definition 18 (Signature arithmetic). *Given $c \in \mathbb{R}_{++}$, a portfolio $\mathbf{X}$ of size k and two positions $\lambda, \lambda' \in \mathbb{R}^k$. The following symbols for lattice instances and signatures are associated:*

symbol	signature
L	$(\mathbf{X}, \lambda)$
L'	$(\mathbf{X}, \lambda')$
cL	$(\mathbf{X}, c\lambda)$
$-L$	$(\mathbf{X}, -\lambda)$
$L + L'$	$(\mathbf{X}, \lambda + \lambda')$

Proposition 5 (Sub-additivity). *Given $c \in \mathbb{R}_{++}$, a portfolio $\mathbf{X}$ of size k and two orthogonal positions λ and λ', i.e., $\lambda_n \lambda'_n = 0$ for $1 \le n \le k$. Let $\hat{F}$ be the best worst-case process for $(\mathbf{X}, \lambda)$..*

Then, in the notation of Def. 18, the following holds for $0 \le i \le N$:

$$\hat{F}_i(cL) = c\,\hat{F}_i(L)$$
$$\hat{F}_i(L + L') \le \hat{F}_i(L) + \hat{F}_i(L') \tag{8.21}$$
$$\hat{F}_i(L + L') \ge \hat{F}_i(L) - \hat{F}'_i(-L')$$

8.1 Early Exercise Combinations

Proof. We only prove the second inequality, by induction over i. For $i = N$, equality holds in (8.21) throughout, as all instruments mature at time $t = t_N$. So let $i < N$, and assume (8.21) is true for $i+1$:

$$\hat{F}_{i+1}(L + L') \leq \hat{F}_{i+1}(L) + \hat{F}_{i+1}(L') \tag{8.22}$$

Let $M \subseteq \operatorname{am}(\mathbf{X})$ be a subset of American instruments on L (and on L' and $L + L'$, for that matter). If $M = \emptyset$ the residual lattice of $L + L'$ is $L + L'$ itself. In this case, direct application of (8.22) leads to

$$\begin{aligned}
F_i(L + L', \emptyset, \sigma) &= \frac{1}{\beta_i} \mathrm{E}_{Q(\sigma)} \left(\beta_{i+1} \hat{F}_{i+1}(L + L') \mid \mathcal{F}_i \right) \\
&\leq \frac{1}{\beta_i} \mathrm{E}_{Q(\sigma)} \left(\beta_{i+1} \left(\hat{F}_{i+1}(L) + \hat{F}_{i+1}(L') \right) \mid \mathcal{F}_i \right) \\
&= \frac{1}{\beta_i} \mathrm{E}_{Q(\sigma)} \left(\beta_{i+1} \hat{F}_{i+1}(L) \mid \mathcal{F}_i \right) \\
&\quad + \frac{1}{\beta_i} \mathrm{E}_{Q(\sigma)} \left(\beta_{i+1} \hat{F}_{i+1}(L') \mid \mathcal{F}_i \right) \\
&= F_i(L, \emptyset, \sigma) + F_i(L', \emptyset, \sigma)
\end{aligned} \tag{8.23}$$

If $M \neq \emptyset$ we have $(L + L')_{\neg M} \subset L + L'$. (8.22) does not directly validate

$$\hat{F}_{i+1}((L + L')_{\neg M}) \leq \hat{F}_{i+1}(L_{\neg M}) + \hat{F}_{i+1}(L'_{\neg M}) \tag{8.24}$$

but we may hold (8.24) to be true, by nested application of the proposition for a portfolio of smaller size $k' = |(L + L')_{\neg M}| < k$. (For $k' = 0$, equality obviously holds.) Thus,

$$\begin{aligned}
F_i(L + L', M, \sigma) &= \frac{1}{\beta_i} \mathrm{E}_{Q(\sigma)} \left(\beta_{i+1} \hat{F}_{i+1}((L + L')_{\neg M}) \mid \mathcal{F}_i \right) + \text{payoff}(L + L', M) \\
&\leq \frac{1}{\beta_i} \mathrm{E}_{Q(\sigma)} \left(\beta_{i+1} \left(\hat{F}_{i+1}(L_{\neg M}) + \hat{F}_{i+1}(L'_{\neg M}) \right) \mid \mathcal{F}_i \right) \\
&\quad + \text{payoff}(L + L', M) \\
&= \frac{1}{\beta_i} \mathrm{E}_{Q(\sigma)} \left(\beta_{i+1} \hat{F}_{i+1}(L_{\neg M}) \mid \mathcal{F}_i \right) + \text{payoff}(L, M) \\
&\quad + \frac{1}{\beta_i} \mathrm{E}_{Q(\sigma)} \left(\beta_{i+1} \hat{F}_{i+1}(L'_{\neg M}) \mid \mathcal{F}_i \right) + \text{payoff}(L', M) \\
&= F_i(L, M, \sigma) + F_i(L', M, \sigma)
\end{aligned} \tag{8.25}$$

and sub-additivity is shown for the local fixation. Since this is true for all $\sigma \in \mathcal{C}$, we get

$$\begin{aligned}
\sup_{\sigma \in \mathcal{C}} F_i(L + L', M, \sigma) &\leq \sup_{\sigma \in \mathcal{C}} \left[F_i(L, M, \sigma) + F_i(L', M, \sigma) \right] \\
&\leq \sup_{\sigma \in \mathcal{C}} F_i(L, M, \sigma) + \sup_{\sigma' \in \mathcal{C}} F_i(L', M, \sigma')
\end{aligned} \tag{8.26}$$

or
$$\hat{F}_i(L+L', M) \leq \hat{F}_i(L, M) + \hat{F}_i(L', M) \tag{8.27}$$

Thus, reintroducing local uncertainty does not violate sub-additivity.

The remainder of the proof is concerned with retaining sub-additivity through the application of the minimum and maximum operators. Just as in (8.7), define

$$\begin{aligned}
A(L) &= \text{long}\,(\mathbf{X}, \lambda) & B(L) &= \text{short}\,(\mathbf{X}, \lambda) \\
A(L') &= \text{long}\,(\mathbf{X}, \lambda_{L'}) & B(L') &= \text{short}\,(\mathbf{X}, \lambda_{L'}) \\
A(L+L') &= \text{long}\,(\mathbf{X}, \lambda_{L+L'}) & B(L+L') &= \text{short}\,(\mathbf{X}, \lambda_{L+L'})
\end{aligned} \tag{8.28}$$

The orthogonality of λ and λ' and Def. 10 imply that

$$\begin{aligned}
A(L+L') &= A(L) \cup A(L') \\
B(L+L') &= B(L) \cup B(L')
\end{aligned} \tag{8.29}$$

where the union is direct, i.e. $A(L) \cap A(L') = \emptyset$ and $B(L) \cap B(L') = \emptyset$.

Now partition $M = A \cup B$ with $A \subseteq A(L+L')$ and $B = B(L+L')$. Then

$$\begin{aligned}
\text{payoff}\,(L, M) &= \text{payoff}\,(L, A \cup B) \\
&= \text{payoff}\,(L, (A \cap A(L)) \cup (B \cap B(L))) \\
\text{payoff}\,(L', M) &= \text{payoff}\,(L', A \cup B) \\
&= \text{payoff}\,(L', (A \cap A(L')) \cup (B \cap B(L')))
\end{aligned} \tag{8.30}$$

and consequently

$$\begin{aligned}
\hat{F}_i(L, M) &= \hat{F}_i(L, A \cup B) \\
&= \hat{F}_i(L, (A \cap A(L)) \cup (B \cap B(L))) \\
\hat{F}_i(L', M) &= \hat{F}_i(L', A \cup B) \\
&= \hat{F}_i(L', (A \cap A(L')) \cup (B \cap B(L')))
\end{aligned} \tag{8.31}$$

Although we do not show this in every detail, (8.30) and (8.31) are easy to validate, since $A \setminus A(L)$ and $B \setminus B(L)$ respectively $A \setminus A(L')$ and $B \setminus B(L')$ refer to vanishing positions: the corresponding λ's respectively λ''s are all zero. It is straightforward to equate payoff terms and signatures of lattice instances that differ only on instruments whose position is zero.

Reintroducing local optionality, it follows from (8.27) that

$$\begin{aligned}
\hat{F}_i(L+L') &= \min_{A \subseteq A(L+L')} \max_{B \subseteq B(L+L')} \hat{F}_i(L+L', A \cup B) \\
&\leq \min_{A \subseteq A(L+L')} \max_{B \subseteq B(L+L')} \left(\hat{F}_i(L, A \cup B) + \hat{F}_i(L', A \cup B) \right)
\end{aligned} \tag{8.32}$$

With (8.31),

8.1 Early Exercise Combinations

$$\min_{A \subseteq A(L+L')} \max_{B \subseteq B(L+L')} \left(\hat{F}_i(L, A \cup B) + \hat{F}_i(L', A \cup B) \right)$$

$$= \min_{A \subseteq A(L+L')} \max_{B \subseteq B(L+L')} \left(\hat{F}_i(L, (A \cap A(L)) \cup (B \cap B(L))) \right.$$
$$\left. + \hat{F}_i(L', (A \cap A(L')) \cup (B \cap B(L'))) \right) \quad (8.33)$$

$$\leq \min_{A \subseteq A(L+L')} \left(\max_{B \subseteq B(L+L')} \hat{F}_i(L, (A \cap A(L)) \cup (B \cap B(L))) \right.$$
$$\left. + \max_{B \subseteq B(L+L')} \hat{F}_i(L', (A \cap A(L')) \cup (B \cap B(L'))) \right)$$

Some of the candidate sets searched by the maximum operators can be dropped, and the candidate sets for the minimum operator can be partitioned:

$$\min_{A \subseteq A(L+L')} \left(\max_{B \subseteq B(L+L')} \hat{F}_i(L, (A \cap A(L)) \cup (B \cap B(L))) \right.$$
$$\left. + \max_{B \subseteq B(L+L')} \hat{F}_i(L', (A \cap A(L')) \cup (B \cap B(L'))) \right)$$

$$= \min_{A \subseteq A(L+L')} \left(\max_{B \subseteq B(L)} \hat{F}_i(L, (A \cap A(L)) \cup B) \right.$$
$$\left. + \max_{B \subseteq B(L')} \hat{F}_i(L', (A \cap A(L')) \cup B) \right)$$

$$= \min_{A_1 \subseteq A(L)} \min_{A_2 \subseteq A(L')} \left(\max_{B \subseteq B(L)} \hat{F}_i(L, A_1 \cup B) + \max_{B \subseteq B(L')} \hat{F}_i(L', A_2 \cup B) \right)$$
$$(8.34)$$

Rearranging terms yields

$$\min_{A_1 \subseteq A(L)} \min_{A_2 \subseteq A(L')} \left(\max_{B \subseteq B(L)} \hat{F}_i(L, A_1 \cup B) + \max_{B \subseteq B(L')} \hat{F}_i(L', A_2 \cup B) \right)$$

$$= \min_{A_1 \subseteq A(L)} \left(\max_{B \subseteq B(L)} \hat{F}_i(L, A_1 \cup B) + \min_{A_2 \subseteq A(L')} \max_{B \subseteq B(L')} \hat{F}_i(L', A_2 \cup B) \right)$$

$$= \min_{A_1 \subseteq A(L)} \max_{B \subseteq B(L)} \hat{F}_i(L, A_1 \cup B) + \min_{A_2 \subseteq A(L')} \max_{B \subseteq B(L')} \hat{F}_i(L', A_2 \cup B)$$
$$(8.35)$$

Since, by definition,

$$\min_{A_1 \subseteq A(L)} \max_{B \subseteq B(L)} \hat{F}_i(L, A_1 \cup B) = \hat{F}_i(L) \quad (8.36)$$

and

$$\min_{A_2 \subseteq A(L')} \max_{B \subseteq B(L')} \hat{F}_i(L', A_2 \cup B) = \hat{F}_i(L') \quad (8.37)$$

we conclude from the sequence (8.32) through (8.35) that

$$\hat{F}_i(L + L') \leq \hat{F}_i(L) + \hat{F}_i(L') \tag{8.38}$$

This completes the induction step and the proof. □

It should be obvious that the assertions of Prop. 5 also hold simultaneously for all partial portfolios $(\mathbf{X}', \lambda') \subseteq (\mathbf{X}, \lambda)$.

In the proposition, the portfolio $\mathbf{X}$ is split into halves $(X_n \mid \lambda_n \neq 0)^T$ and $(X_n \mid \lambda_n = 0)^T$. A different formulation uses two partial portfolios $(\mathbf{X}_i, \lambda_i) \subseteq (\mathbf{X}, \lambda)$, $i = 1, 2$, that do not overlap: $(\mathbf{X}_i, \lambda_i) = (\text{select}(\mathbf{X}, M_i), \text{select}(\lambda, M_i))$ with $M_1 \cap M_2 = \emptyset$ and $M_1 \cup M_2 = \{1, \ldots, |\mathbf{X}|\}$. L_1, L_2 and $-L_2$ being the lattice instances with signatures $(\mathbf{X}_1, \lambda_1)$, $(\mathbf{X}_2, \lambda_2)$ and $(\mathbf{X}_2, -\lambda_2)$, respectively, (8.21) reads

$$\begin{aligned} \hat{F}_i(L) &\leq \hat{F}_i(L_1) + \hat{F}_i(L_2) \\ \hat{F}_i(L) &\geq \hat{F}_i(L_1) - \hat{F}'_i(-L_2) \end{aligned} \tag{8.39}$$

We will refer to the assertions of the proposition in either form, depending on the context.

Implementation In principle, we have already seen in Fig. 8.1 how the best worst-case process for $(\mathbf{X}, \lambda)$ can be implemented by applying dynamic programming principles locally. The variables $\hat{V}(j, i; L)$ are discrete approximizations of the values $\hat{F}_i(L \mid S_i = s_j)$ of the best worst-case process. The results achieved so far motivate the algorithm in Fig. 8.4 to compute the "suitable early exercise combination" mentioned in the caption of Fig. 8.1.

8.2 Speedup Techniques

The term

$$\hat{F}_i(L) = \min_{A \subseteq A(L)} \max_{B \subseteq B(L)} \hat{F}_i(L, A \cup B) \tag{8.40}$$

has $2^{|A(L)|+|B(L)|}$ subexpressions. If corresponds to step 3b in Fig. 8.4. If the signature of L is $(\mathbf{X}, \lambda)$ and there are $n \leq k$ American instruments in $\mathbf{X}$, the running time of the worst-case pricer becomes $O(2^n)$, which is quite unacceptable.

In this section, we explore two ways of improving this performance impasse:

1. Sometimes it can be said with certainty that a particular instrument X_n on a lattice instance L must or must not be exercised, regardless of the remaining position. Only where such certainty cannot be gained is it necessary to consider both possibilities in concert with all other instruments.

Input: Lattice instance L with signature $(\mathbf{X}, \lambda)$, time i
Output: $\hat{V}(j, i; L)$ for $D \leq j \leq U$

1. Repeat for all spatial levels $D \leq j \leq U$, possibly with the algorithm in Fig. 5.5:
   a) Set $\hat{V}(j, i; L, \emptyset) \doteq \sup_{\sigma \in C} \frac{1}{\beta_i} \mathrm{E}_{Q(\sigma)} \left(\beta_{i+1} \hat{F}_{i+1}(L) \mid S_j = s_i \right)$
   b) Set the gradient $\hat{v}_n(j, i; L, \emptyset)$ accordingly, for $1 \leq n \leq |L|$:

$$\hat{v}_n(j, i; L, \emptyset) = \frac{\partial}{\partial \lambda_n} \hat{V}(j, i; L, \emptyset)$$

2. Set $A(L) = \text{long}\,(\mathbf{X}, \lambda)$ and $B(L) = \text{short}\,(\mathbf{X}, \lambda)$
3. Solve the local minmax problem:
   a) For all $\emptyset \neq M \subseteq A(L) \cup B(L)$, check whether $\hat{V}(j, i; L', \emptyset)$ has already been computed, where the signature of L' is $(\text{select}\,(\mathbf{X}, M), \text{select}\,(\lambda, M))$. If not, interrupt and recurse
   b) Find $\hat{A} \subseteq A(L)$ and $\hat{B} \subseteq B(L)$ such that

$$\hat{V}(j, i; L, \hat{A} \cup \hat{B}) = \min_{A \subseteq A(L)} \max_{B \subseteq B(L)} \hat{V}(j, i; L, A \cup B)$$

   and set $M = \hat{A} \cup \hat{B}$, $\hat{\mathbf{X}} = \text{select}\,(\mathbf{X}, M)$, and $\hat{\lambda} = \text{select}\,(\lambda, M)$
   c) If $M = \emptyset$ set $\hat{V}(j, i; L) = \hat{V}(j, i; L, \emptyset)$ and $\hat{v}_n(j, i; L) = \hat{v}_n(j, i; L, \emptyset)$, $1 \leq n \leq |L|$. Otherwise let $\hat{L}$ be the lattice instance with signature $(\hat{\mathbf{X}}, \hat{\lambda})$, set $\hat{V}(j, i; L) = \hat{V}(j, i; \hat{L}) + \text{payoff}\,(L, M)$, and copy $\hat{v}_n(j, i; L)$ from $\hat{L}$ for $n \in \{1, \ldots, |L|\} \setminus M$, after proper reindexing. For all other n, set $\hat{v}_n(j, i; L) = X_n$

Fig. 8.4. The algorithm to track the best worst-case process on the lattice. In a real implementation, step 1 is one round in an explicit or mixed explicit/implicit finite difference scheme, based on PDE's of type (4.9). Step 2 offers potential for improvement. The temporary vector $\hat{V}(\cdot, i; L, \emptyset)$ is the discrete version of $\hat{F}(L, \emptyset)$

2. In the linear case the simple cutoff rule

$$X_n \geq \hat{v}_n(i, j; L, \emptyset) \tag{8.41}$$

comparing the potential payoff with the prospective future profit determines the early exercise boundary. This formula is no longer true in the nonlinear case, but it might well be useful as heuristic.

In both cases, space is partitioned onto three regions.

Definition 19 (Corridor of uncertainty). *Let L be a lattice instance with signature $(\mathbf{X}, \lambda)$. Choose $n \in \text{am}\,(\mathbf{X})$. Let $(j, i; L)$ be a node instance.*

If early exercise of X_n is a priori not pursued at $(j, i; L)$ then $(j, i; L)$ belongs to the continuation region *of X_n.*

If early exercise of X_n is a priori opted for at $(j, i; L)$ then $(j, i; L)$ belongs to the exercise region *of X_n.*

92   8 Algorithms for American Options

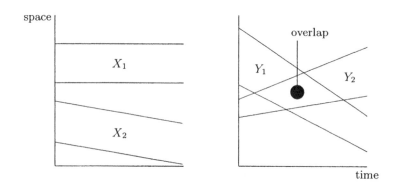

Fig. 8.5. Shown on the left side are the non-overlapping corridors of uncertainty for two American options X_1 and X_2. Within each corridor 2 early exercise alternatives must be considered: execise X_1 resp. X_2 versus don't exercise X_1 resp. X_2. On the right side, the corridors of uncertainty for the American options Y_1 and Y_2 overlap. In the overlap region, 4 early exercise alternatives must be considered. (This picture is conceptual. For an actual example, see Fig. 8.8)

In early exercise is a priori neither avoided nor opted for at $(j, i; L)$ then $(j, i; L)$ belongs to the corridor of uncertainty of X_n. In its corridor of uncertainty, X_n contributes to the exponential complexity of (8.40).

Notice that the terminology is operational: the continuation region of X_n is not the region in which not exercising X_n is optimal in the sense of the minmax formulation. Rather, continuation region, exercise region and corridor of uncertainty are established externally; then (8.40) is applied at each node instance (j, i, L) that belongs to the corridor of uncertainty of at least one instrument.

The computational complexity is still exponential in n where n corridors of uncertainty overlap. Figure 8.5 shows cases with non-overlapping and overlapping corridors of uncertainty.

It is crucial to keep the corridor of uncertainty as small as possible. The first speedup approach uses a worst-case and best-case price band for each individual American instrument to estimate the corridor of uncertainty. This approach never misses the correct combination $(\hat{A}, \hat{B})$ in step 3b of Fig. 8.4 and is thus equivalent.

The second speedup approach collapses the corridor of uncertainty. The formula (8.41) is used to separate space into regions of continuation and exercise, respectively. The corridor of uncertainty is empty. This approach is no longer guaranteed to be correct; it is a heuristic. It handles American options much like barrier options. For instance, it can be made to process

2∗ Partition long $(\mathbf{X}, \lambda) = A_C \cup A_U \cup A_E$ and short $(\mathbf{X}, \lambda) = B_C \cup B_U \cup B_E$, where the subscripts C, U and E stand for <u>c</u>ontinuation region, <u>u</u>ncertainty corridor and <u>e</u>xercise region, respectively

3∗ Solve the local minmax problem:
   a) For all $M \subseteq A_U \cup B_U$, interrupt and recurse if $\hat{V}(j, i; L', \emptyset)$ has not been computed yet. Here, the signature of L' is

   $$(\text{select}(\mathbf{X}, M \cup A_E \cup B_E), \text{select}(\lambda, M \cup A_E \cup B_E))$$

   b) Find $\hat{A} \subseteq A_U$ and $\hat{B} \subseteq B_U$ such that

   $$\hat{V}(j, i; L, \hat{A} \cup \hat{B} \cup A_E \cup B_E)$$
   $$= \min_{A \subseteq A_U} \max_{B \subseteq B_U} \hat{V}(j, i; L, A \cup B \cup A_E \cup B_E)$$

   and set $M = \hat{A} \cup \hat{B} \cup A_E \cup B_E$, $\hat{\mathbf{X}} = \text{select}(\mathbf{X}, M)$, and $\hat{\lambda} = \text{select}(\lambda, M)$
   c) ... (just as step 3c in Fig. 8.4)

Fig. 8.6. Steps 2∗ and 3∗ replace steps 2 and 3 in Fig. 8.4. Step 2∗ is still generic: it does not provide guidelines as to how to partition the long and short positions. Sections 8.2.1 and 8.2.2 fill in the details

---

barrier options with irregular barriers, by replacing (8.41) with $s_j \geq U(X_n, i)$ where U maps to a time-dependent up-and-out barrier for X_n.

These techniques require a refinement of steps 2 and 3 in Fig. 8.4. A general template of the necessary changes is offered in Fig. 8.6.

8.2.1 Maintaining the Corridor of Uncertainty

So far we have created partial portfolios $(\mathbf{X}', \lambda') \subseteq (\mathbf{X}, \lambda)$ only when they were necessary as sources of boundary values. In this section we shall see how the separate computation of best worst-case prices for $(X_n, 1)$ and $(X_n, -1)$, $1 \leq n \leq k$, can help to eliminate early exercise combinations without sacrificing correctness. (The notation $(X_n, \pm 1)$ is used as shorthand for the vector pair $((X_n), (\pm 1))$ throughout this section and the next.)

Proposition 6 (Corridor of uncertainty I). *Let L be a lattice instance of size k with signature $(\mathbf{X}, \lambda)$, and let $\hat{F}$ be the best worst-case price process. For $n \in \text{am}(\mathbf{X})$, let L^U be the lattice instance with signature $(X_n, 1)$, and let L_D be the lattice instance with signature $(X_n, -1)$. Then*

$$-\hat{F}_i(L_D) \leq \hat{F}_i(L_U) \tag{8.42}$$

for $0 \leq i \leq N$.

Proof. By induction over i. For $i = N$ we have equality, as X_n is always exercised at the maturity date t_N, and

8 Algorithms for American Options

$$-\hat{F}_N(L_D) = \hat{F}_N(L_U) = X_N \tag{8.43}$$

For $i < N$, by unrolling Defs. 13, 14, 15 and 16, we get

$$\begin{aligned}-\hat{F}_i(L_D) &= -\min\left\{-X_n, \sup_{\sigma\in\mathcal{C}}\frac{1}{\beta_i}\mathbb{E}_{Q(\sigma)}\left(\beta_{i+1}\hat{F}_{i+1}(L_D)\mid\mathcal{F}_i\right)\right\}\\ &= \max\left\{X_n, -\sup_{\sigma\in\mathcal{C}}\frac{1}{\beta_i}\mathbb{E}_{Q(\sigma)}\left(\beta_{i+1}\hat{F}_{i+1}(L_D)\mid\mathcal{F}_i\right)\right\}\end{aligned} \tag{8.44}$$

Using the induction hypothesis and linearity of expectation,

$$\begin{aligned}&-\sup_{\sigma\in\mathcal{C}}\frac{1}{\beta_i}\mathbb{E}_{Q(\sigma)}\left(\beta_{i+1}\hat{F}_{i+1}(L_D)\mid\mathcal{F}_i\right)\\ &= \inf_{\sigma\in\mathcal{C}}\frac{1}{\beta_i}\mathbb{E}_{Q(\sigma)}\left(-\beta_{i+1}\hat{F}_{i+1}(L_D)\mid\mathcal{F}_i\right)\\ &\leq \inf_{\sigma\in\mathcal{C}}\frac{1}{\beta_i}\mathbb{E}_{Q(\sigma)}\left(\beta_{i+1}\hat{F}_{i+1}(L^U)\mid\mathcal{F}_i\right)\\ &\leq \sup_{\sigma\in\mathcal{C}}\frac{1}{\beta_i}\mathbb{E}_{Q(\sigma)}\left(\beta_{i+1}\hat{F}_{i+1}(L^U)\mid\mathcal{F}_i\right)\end{aligned} \tag{8.45}$$

Together,

$$-\hat{F}_i(L_D) \leq \max\left\{X_n, \sup_{\sigma\in\mathcal{C}}\frac{1}{\beta_i}\mathbb{E}_{Q(\sigma)}\left(\beta_{i+1}\hat{F}_{i+1}(L^U)\mid\mathcal{F}_i\right)\right\} = \hat{F}_i(L^U) \tag{8.46}$$

$\square$

Proposition 6 shows that $-\hat{F}(L_D)$ and $\hat{F}(L^U)$ span a non-empty corridor between them. The following proposition shows that this corridor can be used to separate the continuation and exercise regions.

Proposition 7 (Corridor of uncertainty II). *Given a lattice instance L of size k with signature $(\mathbf{X}, \lambda)$. Let $\hat{F}$ be the best worst-case process for $(\mathbf{X}, \lambda)$. Choose $n \in \text{am}(\mathbf{X})$. Set $A(L) = \text{long}(\mathbf{X}, \lambda)$ and $B(L) = \text{short}(\mathbf{X}, \lambda)$. Let $A'(L) = A(L) \setminus \{n\}$ and $B'(L) = B(L) \setminus \{n\}$ be their reduced versions. Let L^U be the lattice instance with signature $(X_n, 1)$, and let L_D be the lattice instance with signature $(X_n, -1)$.*
If $\hat{F}_i(L^U) \leq X_n$ then

$$\hat{F}_i(L) = \min_{A\subseteq A'(L)}\max_{B\subseteq B'(L)}\hat{F}_i(L, A\cup B\cup\{n\}) \tag{8.47}$$

i.e. X_n is exercised for sure.
If, on the other hand, $-\hat{F}_i(L_D) > X_n$ then

$$\hat{F}_i(L) = \min_{A\subseteq A'(L)}\max_{B\subseteq B'(L)}\hat{F}_i(L, A\cup B) \tag{8.48}$$

i.e. X_n is not exercised.

Proof. The proof uses Prop. 5. Set $M = \{1, \ldots, n-1, n+1, \ldots, k\}$ and

$$(\mathbf{X}', \lambda') = (\text{select}(\mathbf{X}, M), \text{select}(\lambda, M)) \tag{8.49}$$

Let L' be the lattice instance with signature $(\mathbf{X}', \lambda')$, and notice that

$$\hat{F}_i(L') = \min_{A \subseteq A'(L)} \max_{B \subseteq B'(L)} \hat{F}_i(L', A \cup B) \tag{8.50}$$

Let L^+ be the lattice instance with signature (X_n, λ_n), and let L^- be the lattice instance with signature $(X_n, -\lambda_n)$.

Case 1 Assume $\hat{F}_i(L^U) \leq X_n$ and $\lambda_n > 0$, i.e. $n \in B(L)$ and $B'(L) \subset B(L)$. Then $\lambda_n \hat{F}(L^U) = \hat{F}(L^+)$ by the first assertion of Prop. 5. Furthermore, by the second assertion of Prop. 5,

$$\begin{aligned}
\hat{F}_i(L) &\leq \hat{F}_i(L^+) + \hat{F}_i(L') \\
&= \lambda_n \hat{F}_i(L^U) + \hat{F}_i(L') \\
&\leq \lambda_n X_n + \hat{F}_i(L') \\
&= \lambda_n X_n + \min_{A \subseteq A'(L)} \max_{B \subseteq B'(L)} \hat{F}_i(L', A \cup B) \\
&= \min_{A \subseteq A'(L)} \max_{B \subseteq B'(L)} \hat{F}_i(L, A \cup B \cup \{n\})
\end{aligned} \tag{8.51}$$

The last transformation follows because a) $\lambda_n X_n$ can be pulled inside the payoff term of $\hat{F}_i(L', A \cup B)$, and b) the residual lattice instances of L with respect to $A \cup B \cup \{n\}$ and L' with respect to $A \cup B$ are identical.

For fixed A we have $\max_{B \subseteq B(L)} \hat{F}_i(L, A \cup B) \geq \max_{B \subseteq B'(L)} \hat{F}_i(L, A \cup B \cup \{n\})$, since in the latter term the maximum is taken over less values. Since $A(L) = A'(L)$,

$$\begin{aligned}
\hat{F}_i(L) &= \min_{A \subseteq A(L)} \max_{B \subseteq B(L)} \hat{F}_i(L, A \cup B) \\
&\geq \min_{A \subseteq A(L)} \max_{B \subseteq B'(L)} \hat{F}_i(L, A \cup B \cup \{n\}) \\
&= \min_{A \subseteq A'(L)} \max_{B \subseteq B'(L)} \hat{F}_i(L, A \cup B \cup \{n\})
\end{aligned} \tag{8.52}$$

(8.51) and (8.52) together prove (8.47).

Case 2 If $\lambda_n < 0$, i.e. $n \in A(L)$ and $A'(L) \subset A(L)$, we reason similarly. By the first assertion of Prop. 5, $-\lambda_n \hat{F}(L^U) = \hat{F}(L^-)$. By the third assertion of Prop. 7,

$$\hat{F}_i(L) \geq -\hat{F}_i(L^-) + \hat{F}_i(L')$$
$$= -\left(-\lambda_n \hat{F}_i(L^U)\right) + \hat{F}_i(L')$$
$$\geq \lambda_n X_n + \hat{F}_i(L') \tag{8.53}$$
$$= \lambda_n X_n + \min_{A \subseteq A'(L)} \max_{B \subseteq B'(L)} \hat{F}_i(L', A \cup B)$$
$$= \min_{A \subseteq A'(L)} \max_{B \subseteq B'(L)} \hat{F}_i(L, A \cup B \cup \{n\})$$

In the other direction we use $B(L) = B'(L)$ and get

$$\hat{F}_i(L) = \min_{A \subseteq A(L)} \max_{B \subseteq B(L)} \hat{F}_i(L, A \cup B)$$
$$\leq \min_{A \subseteq A'(L)} \max_{B \subseteq B(L)} \hat{F}_i(L, A \cup B \cup \{n\}) \tag{8.54}$$
$$= \min_{A \subseteq A'(L)} \max_{B \subseteq B'(L)} \hat{F}_i(L, A \cup B \cup \{n\})$$

Again, (8.47) follows readily.

Case 3 Now assume $-\hat{F}_i(L_D) > X_n$ and $\lambda_n > 0$, i.e. $n \in B(L)$, $B'(L) \subset B(L)$. Then $\lambda_n \hat{F}(L_D) = \hat{F}(L^-)$ by the first assertion of Prop. 5. With the third assertion of the proposition and $A(L) = A(L')$,

$$\hat{F}_i(L) \geq -\hat{F}_i(L^-) + \hat{F}_i(L')$$
$$= -\lambda_n \hat{F}_i(L_D) + \hat{F}_i(L')$$
$$> \lambda_n X_n + \hat{F}_i(L') \tag{8.55}$$
$$= \min_{A \subseteq A'(L)} \max_{B \subseteq B'(L)} \hat{F}_i(L, A \cup B \cup \{n\})$$

Choose $\hat{A} \subseteq A(L)$ and $\hat{B} \subseteq B(L)$ such that $\hat{F}_i(L) = \hat{F}_i(L, \hat{A} \cup \hat{B})$. If $n \in \hat{B}$ the strict inequalitiy in (8.55) leads to a contradiction. Thus, $n \notin \hat{B}$. Therefore, $\hat{B} \subseteq B(L) \setminus \{n\} = B'(L)$, and

$$\hat{F}_i(L) = \min_{A \subseteq A'(L)} \max_{B \subseteq B'(L)} \hat{F}_i(L, A \cup B) \tag{8.56}$$

Case 4 If $\lambda_n < 0$, i.e. $n \in A(L)$ and $A'(L) \subset A(L)$, we find that $-\lambda_n \hat{F}(F_D) = \hat{F}(L^+)$ by the first assertion of Prop. 5. The second assertion of Prop. 5 and $B(L) = B(L')$ imply

$$\hat{F}_i(L) \leq \hat{F}_i(L^+) + \hat{F}_i(L')$$
$$= -\lambda_n \hat{F}_i(L_D) + \hat{F}_i(L')$$
$$< \lambda_n X_n + \hat{F}_i(L') \tag{8.57}$$
$$= \min_{A \subseteq A'(L)} \max_{B \subseteq B'(L)} \hat{F}_i(L, A \cup B \cup \{n\})$$

2** Set

$$A_C = \left\{ n \in \text{long}(\mathbf{X}, \lambda) \mid -\hat{V}(j, i; L_D^n) > X_n \right\}$$
$$A_F = \left\{ n \in \text{long}(\mathbf{X}, \lambda) \mid \hat{V}(j, i; L_n^U) \leq X_n \right\}$$
$$A_M = \text{long}(\mathbf{X}, \lambda) \setminus (A_C \cup A_F)$$

and

$$B_C = \left\{ n \in \text{short}(\mathbf{X}, \lambda) \mid -\hat{V}(j, i; L_D^n) > X_n \right\}$$
$$B_F = \left\{ n \in \text{short}(\mathbf{X}, \lambda) \mid \hat{V}(j, i; L_n^U) \leq X_n \right\}$$
$$B_M = \text{short}(\mathbf{X}, \lambda) \setminus (B_C \cup B_F)$$

Fig. 8.7. An elaboration of step 2∗ in Fig. 8.6, based on Prop. 7. L_n^U is the lattice instance with signature $(X_n, 1)$, and L_D^n is the lattice instance with signature $(X_n, -1)$, insofar X_n is American. Proposition 6 guarantees that $A_C \cap A_F = B_C \cap B_F = \emptyset$

The argument that concluded case 3 works in this case as well, and thus (8.47) is shown. This completes the proof. □

Figure 8.7 shows how Prop. 7 can be used to initialize the corridor of uncertainty in step 2∗ in Fig. 8.6. The discrete formulation in terms of node instances is straightforward. Note that up to $2k$ additional lattice instances must be maintained. The technique is thus not entirely overhead free, but the overhead is linear in the number of American options. (In fact, it can be shown that the total number of lattice instances is bounded from above by $2^k + k - 1$. The exhaustive set of 2^k partial portfolios includes already k of the additional singleton lattice instances, and "-1" comes from the fact that the empty partial portfolio need not be carried on a lattice instance at all.)

Figure 8.8 shows the location of the corridor of uncertainty for three 30-day American puts with strikes 90, 100 and 110. Under a scenario in which the volatility stays within 10 and 20% the corridors do not overlap. If L is a lattice instance with signature $(\mathbf{X}, \lambda)$, and the three American puts are part of $\mathbf{X}$ (but no other American instruments are), then $|A_M| + |B_M| = 1$ in step 2∗∗, Fig. 8.7, in each corridor, and 0 otherwise. Under a 10–35% scenario, the corridors for the puts with strikes 90 and 100, and the corridors for the puts with strikes 100 and 110 overlap, respectively. Here, $|A_M| + |B_M| = 2$ in the shaded regions, leading to 4 combinations in the minmax term in step 3∗b, Fig. 8.6. Under a 10–50% scenario all corridors overlap, and $|A_M| + |B_M| = 3$ in the shaded region, leading to 8 combinations in the minmax term. The example demonstrates that the corridor of uncertainty is a powerful tool to reduce the combinatorial complexity of the best worst-case pricing problem if

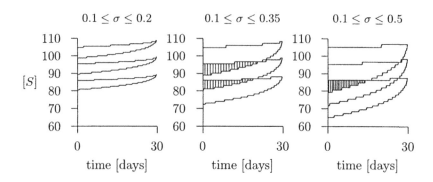

Fig. 8.8. The corridor of uncertainty for three 30-day American puts, with strike 90, 100 and 110, respectively (from bottom to top). The interest rate is $r = 0.03$. The volatility range gets wider from left to right. In the left picture, there is no overlap; in the middle picture, corridors of uncertainty overlap pairwise; in the right picture, all corridors overlap in the shaded region

the volatility range is not too wide. It also shows that the method reverts to exponential complexity if the volatility range is extraordinarily wide. In the next section, a heuristic is presented that tries to alleviate this dependency.

8.2.2 Collapsing the Corridor of Uncertainty

The complexity of best worst-case pricing is potentially exponential if the corridors of uncertainty are nonempty and overlap. To collapse the corridor of uncertainty means to select a definite early exercise boundary, possibly within the corridor, that divides the lattice into continuation and exercise regions for each American option. The way in which this is done in the following paragraphs makes the approach heuristic: it does not guarantee that the resulting early exercise combinations reflect the local best worst-case selections adequately. We present, however, experimental results that show that the discrepancy is negligible in most cases.

The idea is to apply the rule commonly used in linear lattice-based pricing of American options: if the early exercise payoff exceeds the expected future payoff (which includes possible future early exercise), then the expected future payoff is locally replaced by the early exercise payoff. This cut-off rule dynamically assigns each lattice node to either the continuation or exercise region of the lattice.

Under nonlinearity, we do not have an isolated expected future payoff for an American option X_n which is part of a larger portfolio $(\mathbf{X}, \lambda)$. We do have, however, the discrete version of the gradient of the best worst-case value of $(\mathbf{X}, \lambda)$ with respect to the position of X_n, namely

8.2 Speedup Techniques   99

$$\hat{v}_n(j,i;L) \doteq \frac{\partial}{\partial \lambda_n} \hat{F}_i(L \mid S_i = s_j) \qquad (8.58)$$

(Discrete and continuous terms with analogue interpretations are equated with "$\doteq$".) As we will see in the following proposition, this gradient together with λ provides sufficient information to reconstruct the best worst-case value locally.

In this section, we adopt a candidate set

$$\mathcal{C} = \{\sigma \mid \sigma_{\min} \leq \sigma \leq \sigma_{\max}\} \qquad (8.59)$$

with constants $0 < \sigma_{\min} \leq \sigma_{\max}$ for simplicity.

Definition 20 (Local stability). *Let L be a lattice instance with signature $(\mathbf{X}, \lambda)$. Choose $n \in \{1, \ldots, |L|\}$. We say $\hat{F}$ is locally stable with respect to λ_n if*

$$\frac{\partial}{\partial \lambda_n} \hat{F}_i(L) = \frac{\partial}{\partial \lambda_n} \hat{F}_i(L, M) \qquad (8.60)$$

whenever

$$\hat{F}_i(L) = \hat{F}_i(L, M) \qquad (8.61)$$

for some $M \subseteq \{1, \ldots, |L|\}$.

We are not too concerned with the situation in which $\hat{F}$ is not locally stable. In all practical cases, there are only finitely many regions in λ-space with distinct best worst-case exercise combinations at a given node instance (j, i). Finiteness suggests that whenever $\hat{F}_i(\cdot)$ switches from $\hat{F}_i(\cdot, M_1)$ to $\hat{F}_i(\cdot, M_2)$ with $M_1 \neq M_2$, say at $\lambda_n = \alpha$ with all other λ's unchanged, then there are intervals $(\alpha - \epsilon, \alpha)$ and $(\alpha, \alpha + \epsilon)$ in which M_1 respectively M_2 remain the best worst-case exercise combination. If we assume that partial derivatives in λ_n exist for $\hat{F}_i(\cdot)$, then by approaching α from above and below it follows that $\hat{F}$ is locally stable with respect to λ_n.

The subsequent propositions are meant to *motivate* heuristic (8.58). We therefore leave it at this rather informal argument and, by assuming the existence of partial derivatives in λ, at the same time implicitly assert that Def. 20 applies to $\hat{F}$.

Proposition 8. *Let L be a lattice instance of size k with signature $(\mathbf{X}, \lambda)$. Assume the partial derivatives with respect to λ of the best worst-case price process $\hat{F}$ exist. Then the following identity holds:*

$$\sum_{n=1}^{k} \lambda_n \frac{\partial}{\partial \lambda_n} \hat{F}_i(L) = \hat{F}_i(L) \qquad (8.62)$$

8 Algorithms for American Options

Proof. By induction over i. For $i = N$ we have

$$\hat{F}_N(L) = \text{payoff}(L, \{1, \ldots, k\}) = \lambda \cdot \mathbf{X} \tag{8.63}$$

which, as linear combination of individual payoffs, obviously satisfies (8.62).

Now assume $i < N$. By induction hypothesis,

$$\sum_{n=1}^{k} \lambda_n \frac{\partial}{\partial \lambda_n} \hat{F}_{i+1}(L) = \hat{F}_{i+1}(L) \tag{8.64}$$

Case 1 Assume that $\hat{F}_i(L) = \hat{F}_i(L, \emptyset)$, i.e. no instruments are exercised in the best worst-case. Now recall that $\hat{F}_i(L, \emptyset \mid S_i = s_j) = \hat{f}(s_j, i)$, where $\hat{f}$ satisfies a PDE of type (4.9) between times i and $i+1$ (during which interval no exercise takes place), and the boundary value is given by $\hat{F}_{i+1}(L)$. The worst-case spot volatility is determined by the spot convexity $x = \frac{\partial^2 \hat{f}}{\partial S^2}$ and the function

$$\Sigma(x) = \begin{cases} \sigma_{\max} & \text{if } x \geq 0 \\ \sigma_{\min} & \text{if } x < 0 \end{cases} \tag{8.65}$$

In particular,

$$\frac{\partial}{\partial \lambda_n} \Sigma(x) = 0 \tag{8.66}$$

for $x \neq 0$, and $\Sigma(x) x = 0$ for $x = 0$. Differentiating the PDE (4.9) with respect to λ_n shows that $\frac{\partial}{\partial \lambda_n} \hat{f}$ satisfies the same PDE, and so does any linear combination of partial derivatives.

We set the boundary at time $i+1$, and conclude furthermore from (8.64) that the boundary condition is the same for PDE (4.9) and the linear combination of its partial derivatives. Thus, by the uniqueness of the solution of (4.9),

$$\sum_{n=1}^{k} \lambda_n \frac{\partial}{\partial \lambda_n} \hat{F}_i(L, \emptyset) = \hat{F}_i(L, \emptyset) \tag{8.67}$$

This completes the proof for case 1.

Case 2 Now assume $\hat{F}_i(L) = \hat{F}_i(L, M)$ for some $M \neq \emptyset$. Then, with $L' = L_{\neg M}$ being the residual lattice instance,

$$\begin{aligned} \hat{F}_i(L) &= \sup_{\sigma \in C} F_i(L, M, \sigma) \\ &= \left[\sup_{\sigma \in C} F_i(L', \emptyset, \sigma)\right] + \text{payoff}(L, M) \\ &= \hat{F}_i(L', \emptyset) + \text{payoff}(L, M) \end{aligned} \tag{8.68}$$

8.2 Speedup Techniques   101

by the definition of $\hat{F}_i(L, M)$ and $F_i(L, M, \sigma)$ in Defs. 14 and 15, and switching between lattice instances.

Let $k' = |L| - |M| = |L'|$. The induction hypothesis (8.64) makes no statement for the smaller lattice instance L'. We may, however, apply the proposition on L' (inheriting all the premises) and conclude

$$\sum_{n=1}^{k'} (\lambda_{L'})_n \frac{\partial}{\partial (\lambda_{L'})_n} \hat{F}_i(L', \emptyset) = \hat{F}_i(L', \emptyset) \tag{8.69}$$

(A similar argument has been used in Prop. 5. The obvious base case $k' = 0$, or even $k' = 1$, can be worked out directly.)

Let $\{n_1, \ldots, n_{|M|}\}$ be an enumeration of M. As in (8.63),

$$\text{payoff}(L, M) = \sum_{l=1}^{|M|} \lambda_{n_l} \frac{\partial}{\partial \lambda_{n_l}} \text{payoff}(L, M) \tag{8.70}$$

and hence, by adding (8.69) and (8.70) which sum over distinct positions,

$$\sum_{n=1}^{k} \lambda_n \frac{\partial}{\partial \lambda_n} \hat{F}_i(L, M) = \hat{F}_i(L, M) \tag{8.71}$$

This completes the proof.   □

The next two propositions establish upper and lower bounds for the partial derivatives of the best worst-case process.

Proposition 9 (Upper bounds for partial derivatives). *Let L be a lattice instance with signature $(\mathbf{X}, \lambda)$. Assume the partial derivatives of the best worst-case price process $\hat{F}$ in λ exist and are uniformly continuous. Choose $n \in \{1, \ldots, |L|\}$ and let L^U be the lattice instance with signature $(X_n, 1)$. Then*

$$\frac{\partial}{\partial \lambda_n} \hat{F}_i(L) \leq \hat{F}_i(L^U) \tag{8.72}$$

Proof. By induction over i. If $i = N$ all instruments are forcibly exercised, and basic algebra shows that equality holds in (8.72). Thus assume $i < N$, and reason as in the proof of Prop. 8.

Case 1 Assume that $\hat{F}_i(L) = \hat{F}_i(L, \emptyset)$, i.e. no instruments are exercised in the best worst-case. Fix σ. Then, by uniform continuity and the induction hypothesis,

$$\frac{\partial}{\partial \lambda_n} F_i(L, \emptyset, \sigma) = \frac{\partial}{\partial \lambda_n} \frac{1}{\beta_i} \mathrm{E}_{Q(\sigma)} \left(\beta_{i+1} \hat{F}_{i+1}(L) \mid \mathcal{F}_i \right)$$

$$= \frac{1}{\beta_i} \mathrm{E}_{Q(\sigma)} \left(\beta_{i+1} \frac{\partial}{\partial \lambda_n} \hat{F}_{i+1}(L) \mid \mathcal{F}_i \right) \tag{8.73}$$

$$\leq \frac{1}{\beta_i} \mathrm{E}_{Q(\sigma)} \left(\beta_{i+1} \hat{F}_{i+1}(L^U) \mid \mathcal{F}_i \right)$$

8 Algorithms for American Options

The local fixation of $\hat{F}_i(L^U)$ with respect to $\emptyset$ is

$$F_i(L^U, \emptyset, \sigma) = \frac{1}{\beta_i} E_{Q(\sigma)}\left(\beta_{i+1} \hat{F}_{i+1}(L^U) \mid \mathcal{F}_i\right) \tag{8.74}$$

and therefore

$$\hat{F}_i(L^U) = \max\left\{X_n, \sup_{\sigma' \in C} F_i(L^U, \emptyset, \sigma')\right\} \tag{8.75}$$
$$\geq F_i(L^U, \emptyset, \sigma)$$

Together, (8.73), (8.74) and (8.75) show that

$$\frac{\partial}{\partial \lambda_n} F_i(L, \emptyset, \sigma) \leq \hat{F}_i(L^U) \tag{8.76}$$

(8.76) is true for every $\sigma \in C$. Now let $\sigma_1, \sigma_2, \ldots$ be a sequence such that

$$\lim_{l \to \infty} F_i(L, \emptyset, \sigma_l) = \hat{F}_i(L, \emptyset) \tag{8.77}$$

Then,

$$\begin{aligned}
\frac{\partial}{\partial \lambda_n} \hat{F}_i(L, \emptyset) &= \frac{\partial}{\partial \lambda_n} \lim_{l \to \infty} F_i(L, \emptyset, \sigma_l) \\
&= \lim_{l \to \infty} \frac{\partial}{\partial \lambda_n} F_i(L, \emptyset, \sigma_l) \\
&\leq \lim_{l \to \infty} \hat{F}_i(L^U) \\
&= \hat{F}_i(L^U)
\end{aligned} \tag{8.78}$$

As $\hat{F}_i(L) = \hat{F}_i(L, \emptyset)$, we conclude

$$\frac{\partial}{\partial \lambda_n} \hat{F}_i(L) \leq \hat{F}_i(L^U) \tag{8.79}$$

Case 2 Now assume $\hat{F}_i(L) = \hat{F}_i(L, M)$ for some $M \neq \emptyset$. Then, with $L' = L_{\neg M}$ being the residual lattice instance,

$$\hat{F}_i(L) = \hat{F}_i(L', \emptyset) + \text{payoff}(L, M) \tag{8.80}$$

as shown before in (8.68). If $n \notin M$ we apply the proposition for $\hat{F}_i(L', \emptyset)$ with smaller lattice instance L'. If $n \in M$ we take from (8.75) that

$$\begin{aligned}
\hat{F}_i(L^U) &\geq X_n \\
&= \frac{\partial}{\partial \lambda_n} \text{payoff}(L, M) \\
&= \frac{\partial}{\partial \lambda_n} \hat{F}_i(L, M)
\end{aligned} \tag{8.81}$$

This completes the proof. □

8.2 Speedup Techniques

The lower bounds for the partial derivatives of the best worst-case prices are weaker:

Proposition 10 (Lower bounds for partial derivatives). *Let L be a lattice instance with signature $(\mathbf{X}, \lambda)$. Assume the partial derivatives of the best worst-case price process $\hat{F}$ in λ exist and are uniformly continuous. Pick $n \in \{1, \ldots |L|\}$. Let L_D^E be the lattice instance with signature $(X_n^E, -1)$, where X_n^E denotes the European version of X_n (see Def. 11). Let $\hat{F}^E$ be the best worst-case price process for $(X_n^E, -1)$. Then*

$$-\hat{F}_i^E(L_D^E) \leq \frac{\partial}{\partial \lambda_n} \hat{F}_i(L) \tag{8.82}$$

Proof. By induction over i. At maturity ($i = N$), all instruments are exercised, and equality holds in (8.82). Therefore assume $i < N$.

Case 1 Assume that $\hat{F}_i(L) = \hat{F}_i(L, \emptyset)$, i.e. no instruments are exercised in the best worst-case. Fix σ. Then, by uniform continuity and the induction hypothesis,

$$\begin{aligned}
\frac{\partial}{\partial \lambda_n} F_i(L, \emptyset, \sigma) &= \frac{\partial}{\partial \lambda_n} \frac{1}{\beta_i} \mathrm{E}_{Q(\sigma)}\left(\beta_{i+1} \hat{F}_{i+1}(L) \mid \mathcal{F}_i\right) \\
&= \frac{1}{\beta_i} \mathrm{E}_{Q(\sigma)}\left(\beta_{i+1} \frac{\partial}{\partial \lambda_n} \hat{F}_{i+1}(L) \mid \mathcal{F}_i\right) \\
&\geq \frac{1}{\beta_i} \mathrm{E}_{Q(\sigma)}\left(\beta_{i+1} \left(-\hat{F}_{i+1}^E(L_D^E)\right) \mid \mathcal{F}_i\right) \\
&= -\frac{1}{\beta_i} \mathrm{E}_{Q(\sigma)}\left(\beta_{i+1} \hat{F}_{i+1}^E(L_D^E) \mid \mathcal{F}_i\right)
\end{aligned} \tag{8.83}$$

Furthermore

$$\begin{aligned}
-\hat{F}_i^E(L_D^E) &= -\sup_{\sigma' \in \mathcal{C}} \frac{1}{\beta_i} \mathrm{E}_{Q(\sigma')}\left(\beta_{i+1} \hat{F}_{i+1}^E(L_D^E) \mid \mathcal{F}_i\right) \\
&= \inf_{\sigma' \in \mathcal{C}} \left[-\frac{1}{\beta_i} \mathrm{E}_{Q(\sigma')}\left(\beta_{i+1} \hat{F}_{i+1}^E(L_D^E) \mid \mathcal{F}_i\right)\right] \\
&\leq -\frac{1}{\beta_i} \mathrm{E}_{Q(\sigma)}\left(\beta_{i+1} \hat{F}_{i+1}^E(L_D^E) \mid \mathcal{F}_i\right)
\end{aligned} \tag{8.84}$$

Together, (8.83) and (8.84) show that, for every $\sigma \in \mathcal{C}$,

$$\frac{\partial}{\partial \lambda_n} F_i(L, \emptyset, \sigma) \geq \hat{F}_i^E(L_D^E) \tag{8.85}$$

Let $\sigma_1, \sigma_2, \ldots$ be a sequence such that

$$\lim_{l \to \infty} F_i(L, \emptyset, \sigma_l) = \hat{F}_i(L, \emptyset) \tag{8.86}$$

2** Set

$$A_C = \{n \in \text{long}(\mathbf{X}, \lambda) \mid \hat{v}_n(j, i; L, \emptyset) > X_n\}$$
$$A_F = \text{long}(\mathbf{X}, \lambda) \setminus A_C$$

and

$$B_C = \{n \in \text{short}(\mathbf{X}, \lambda) \mid \hat{v}_n(j, i; L, \emptyset) > X_n\}$$
$$B_F = \text{short}(\mathbf{X}, \lambda) \setminus B_C$$

Fig. 8.9. An elaboration of step 2* in Fig. 8.6 that uses partial derivatives to estimate the early exercise boundary. The variables $\hat{v}_n(j, i; L, \emptyset)$ have already been computed

Then,

$$\begin{aligned}
\frac{\partial}{\partial \lambda_n} \hat{F}_i(L, \emptyset) &= \frac{\partial}{\partial \lambda_n} \lim_{l \to \infty} F_i(L, \emptyset, \sigma_l) \\
&= \lim_{l \to \infty} \frac{\partial}{\partial \lambda_n} F_i(L, \emptyset, \sigma_l) \\
&\geq \lim_{l \to \infty} \hat{F}_i^E(L_D^E) \\
&= \hat{F}_i^E(L_D^E)
\end{aligned} \qquad (8.87)$$

Finally,

$$\frac{\partial}{\partial \lambda_n} \hat{F}_i(L) \geq \hat{F}_i^E(L_D^E) \qquad (8.88)$$

Case 2 The case $\hat{F}_i(L) = \hat{F}_i(L, M)$ for some $M \neq \emptyset$ is handled just as case 2 in the proof of Prop. 9. □

Propositions 9 and 10 show that $\frac{\partial}{\partial \lambda_n} \hat{F}_i(L)$ lies in the interval

$$\hat{F}_i^E(L_D^E) \leq \frac{\partial}{\partial \lambda_n} \hat{F}_i(L) \leq \hat{F}_i(L^U) \qquad (8.89)$$

for X_n. This band is wider than the corridor of uncertainty $\left[\hat{F}_i(L_D), \hat{F}_i(L^U)\right]$. We were unable to prove that the lower corridor bound is also a lower bound for the partial derivative.

Of course, $\frac{\partial}{\partial \lambda_n} \hat{F}_i(L)$ is not available in a program unless all early exercise combinations have already been examined. This can be avoided by substituting $\frac{\partial}{\partial \lambda_n} \hat{F}_i(L, \emptyset)$ for $\frac{\partial}{\partial \lambda_n} \hat{F}_i(L)$. Although not shown here, the previous results can be extended to (and partially included already) the estimate

$$\hat{F}_i^E(L_D^E, \emptyset) \le \frac{\partial}{\partial \lambda_n} \hat{F}_i(L, \emptyset) \le \hat{F}_i(L^U, \emptyset) \tag{8.90}$$

Figure 8.9 instantiates the algorithm of Fig. 8.6 to collapse the corridor of uncertainty.

8.2.3 Miscellaneous Issues

Sections 8.2.1 and 8.2.2 have presented the big picture. In this section we review some minor or unresolved issues which are interesting purely from a computational point of view. They are of no financial or numerical concern.

Dynamic Maintenance of the Corridor of Uncertainty The algorithm in Fig. 8.7 relies of the existence of $\hat{V}(j, i; L_n^U)$ and $-\hat{V}(j, i; L_D^n)$ to partition long $(\mathbf{X}, \lambda)$ respectively short $(\mathbf{X}, \lambda)$ into A_C, A_F, A_M respectively B_C, B_F, B_M. Here, L_n^U is the lattice instance with signature $(X_n, 1)$, and L_D^n is the lattice instance with signature $(X_n, -1)$.

Depending on the shape of the lattice (box or tree shape?), its position (what is s_0?), the width of the volatility range and the characteristics of the instruments, A_C or A_F respectively B_C or B_F may sometimes be empty, corresponding to the respective boundaries lying outside the region covered by the lattice.

Thus L_n^U or L_D^n may sometimes be superfluous. In order to not maintain lattice instances which are of no use, we employ the dynamic lookup approach that reduces the number of lattice instances carrying partial portfolios in the first place. The recursion that adds lattice instances when needed is activated in step 3∗a in the algorithm in Fig. 8.6.

The idea is to use the partial derivative of $\hat{V}(j, i; L, \emptyset)$ with respect to λ_n to make a first choice. If $X_n < \hat{v}_n(j, i; L)$ then $\hat{V}(j, i; L_n^U) \le X_n$ cannot be the case, due to Prop. 9 that says, in its discrete approximation, $\hat{v}_n(j, i; L) \le \hat{V}(j, i; L_n^U)$. This necessarily implies $x \notin A_F \cup B_F$. Thus, lookup of L_n^U can be avoided in some cases. As $\hat{v}_n(j, i; L)$ is not yet available when the comparison needs to be made, $\hat{v}_n(j, i; L, \emptyset)$ may be used instead.

In the other direction the situation is more subtle. Prop. 10 only states that $-\hat{F}_i^E(L_D^E) \le \frac{\partial}{\partial \lambda_n} \hat{F}_i(L)$ which is too weak to allow conclusions from $X_n > \hat{v}_n(j, i; L)$. However, under the conjecture $-\hat{F}_i(L_D) \le \frac{\partial}{\partial \lambda_n} \hat{F}_i(L)$, the initial comparison with the partial derivative may indeed lead to the avoidance of the L_D^n lookup for some X_n. This strategy is pursued in our implementation.

A careful look at the data in Fig. 8.12 reveals that this avoidance strategy has practical impact. The number of lattice instances reported in the table are smaller than the ones that follow from the schematic view in Fig. 8.14, for $\sigma_{\max} \le 0.4$.

Recursion Leads to Domino Effect There is also the possibility of the recursion in step 3∗a in Fig. 8.6 causing a domino effect that restarts the

B	BC	AB	ABC
			30 ✓, 29 recurse for BC
	30,29 ✓		resume 29
	28–20 ✓		28–20 ✓
	19 recurse for B		
30–19 ✓	resume 191		19 recurse for AB
		30 ✓, 29 recurse for B	
30,29 ✓		resume 29	
28–19 ✓		28–19 ✓	resume 19
18–0 ✓	18–0 ✓	18–0 ✓	18–0 ✓

Fig. 8.10. Rolling back the lattice for a 30-day up-and-out barrier option A, a 25-day vanilla option B and a 20-day down-and-out barrier option C. Numbers indicate time slices i for which values $\hat{V}(\cdot, i, L)$ are being computed, and labels indicate actions triggered due to lookup misses. ("19 recurse for AB", for instance, means that the lattice instance for portfolio AB does not exist or cannot provide the data for the desired time slice $i = 19$.) The computation proceeds row by row, and within rows from left to right columns. The boxes represent the single case in which the creation of a new lattice instance leads to a waste of compute time

rollback of the time slices in the finite difference scheme for many lattice instances. If $\hat{V}(j, i; L, A_1 \cup B_1 \cup A_E \cup B_E)$ is not available in the computation of $\hat{V}(j, i; L)$ for some node instance $(j, i; L)$, then lattice instance L_1 with signature (select $(\mathbf{X}, M_1)$, select (λ, M_1)) needs to be created. Here, $M_1 = \{1, \ldots, |L|\} \setminus \{A_1 \cup B_1 \cup A_E \cup B_E\}$. The finite difference scheme computes $\hat{V}(\cdot, i'; L_1)$ for all $N \geq i' \geq i$ and resumes the computation of $\hat{V}(j, i; L)$.

A memory-aware implementation of the finite difference scheme does not keep all the values of $\hat{V}(\cdot, i'; L_1)$, and the other lattice instances. Rather, data is kept for the current and the previous time slices i' and $i' + 1$, in order to reduce the space complexity for one lattice instance from $O(N \times (U - D))$ to $O(U - D)$. Here, D and U are the spatial levels of the lattice boundaries.

For this reason, subordinate values $\hat{V}(\cdot, i'; L_1, A_2 \cup B_2 \cup A_E \cup B_E)$ required in turn as data for L_1 need not be available, even if the associated lattice instance L_2 exists. Each lattice instance is equipped to provide data for one "current" time slice, and no others.

We have briefly mentioned in the beginning of Sect. 8.2 that the algorithms for American options are applicable to barrier options with regular or irregular barriers as well. The tools developed in Chapter 7, in particular the algorithm in Fig. 7.4 to compute the extension hierarchy, go beyond the general approach of Fig. 8.6 in that they guarantee that $\hat{V}(j, i; L, A_E \cup B_E)$ is always available if the barriers are canonical.

For illustration purposes, we assume that the computation of the extension hierarchy is turned off in the following example shown in Fig. 8.10. The portfolio consists of a 30-day up-and-out barrier option A, a 25-day vanilla option B and a 20-day down-and-out barrier option C. The time step is one day: $dt = 1/365$.

Figure 8.10 monitors the finite difference scheme time slice by time slice. "Recurse" and "resume" labels indicate where recursion is triggered and work is resumed, and for what time slice. Initially, only the lattice instance for the entire portfolio is maintained. The boxed cells represent a situation in which a total restart is required: the lattice instance $L(B)$ for B is carried unimpeded through time slices $30, \ldots, 19$, when the creation of the lattice instance for AB requires access to time 30-values on $L(B)$. These have been discarded long ago, and so $L(B)$ has to be restarted, resulting in double work for 12 time slices on $L(B)$.

In general, if the directed acyclic graph implied by lookup operations, where vertices model lattices instances and edges model data flow, is recombining, then the domino effect may occur. Edges in the graph are created at different times and may connect to vertices whose lattice instances are not synchronized. Singleton partial portfolios are likely to lead to the domino effect, for instance..

The domino effect can have serious consequences for the running time. There are two solutions to this problem:

1. Develop a tool that precomputes an anlogue of the extension hierarchy for American options. It is in principle possible to evaluate all singleton portfolios first and create a data structure with geometric information on overlapping corridors of uncertainty. The resulting extension hierarchy would be exact. This is a preemptive solution.
2. Periodically checkpoint by saving the values $\hat{V}(\cdot, i; L)$ and related information such as the gradient in separate memory space. A restart can then be based on the data collected during the most recent checkpoint. This solution tries to alleviate the effect of a restart.

Neither approach has been implemented in our system. The domino effect plays no role in our laboratory test cases.

Intermediate Results in the Minmax Computation Step 3∗b in Fig. 8.6 requires the computation of

$$V(L) = \min_{A \subseteq A_U} \max_{B \subseteq B_U} \hat{V}(j, i; L, A \cup B \cup A_E \cup B_E) \qquad (8.91)$$

where the existence of $\hat{V}(j, i; L, A \cup B \cup A_E \cup B_E)$ is guaranteed by step 3∗a. Expansion leads to

$$V(L) = \min \left\{ \max\left[\hat{V}(j, i; L, A_E \cup B_E), \max_{\substack{B \subseteq B_U \\ B \neq \emptyset}} \hat{V}(j, i; L, B \cup A_E \cup B_E) \right], \right.$$

$$\left. \min_{\substack{A \subseteq A_U \\ A \neq \emptyset}} \max_{B \subseteq B_U} \hat{V}(j, i; L, A \cup B \cup A_E \cup B_E) \right\} \qquad (8.92)$$

8 Algorithms for American Options

Furthermore,

$$\max_{\substack{B \subseteq B_U \\ B \neq \emptyset}} \hat{V}(j, i; L, B \cup A_E \cup B_E)$$

$$= \max_{n \in B_U} \max_{B \subseteq B_U \setminus \{n\}} \hat{V}(j, i; L, B \cup A_E \cup B_E \cup \{n\}) \quad (8.93)$$

$$= \max_{n \in B_U} \left(\max_{B \subseteq B_U \setminus \{n\}} \hat{V}(j, i; L_n, B \cup A_E \cup B_E) + \lambda_n X_n \right)$$

where the signature of L_n is

$$\bigl(\text{select}\,(\mathbf{X}, \{1, \ldots, n-1, n+1, \ldots, k\}),$$
$$\text{select}\,(\lambda, \{1, \ldots, n-1, n+1, \ldots, k\})\bigr)$$

Here we assume the computation of A_E, B_E, A_U and B_U is independent of the lattice instance: switching to lattice instance L_n must not change these sets, apart from $\{n\}$. In the algorithms presented so far this is indeed so.

Now consider A_U:

$$\min_{\substack{A \subseteq A_U \\ A \neq \emptyset}} \max_{B \subseteq B_U} \hat{V}(j, i; L, A \cup B \cup A_E \cup B_E)$$

$$= \min_{m \in A_U} \min_{A \subseteq A_U \setminus \{m\}} \max_{B \subseteq B_U} \hat{V}(j, i; L, A \cup B \cup A_E \cup B_E \cup \{m\}) \quad (8.94)$$

$$= \min_{m \in A_U} \left(\hat{V}(j, i; L_m) + \lambda_m X_m \right)$$

Together,

$$V(L) = \min \left\{ \max_{n \in B_U} \left(\max_{B \subseteq B_U \setminus \{n\}} \hat{V}(j, i; L_n, B \cup A_E \cup B_E) + \lambda_n X_n \right), \right.$$
$$\left. \min_{m \in A_U} \left(\hat{V}(j, i; L_m) + \lambda_m X_m \right) \right\} \quad (8.95)$$

Thanks to step 3∗a, $\hat{V}(j, i; L_m)$ is avaliable. It is reasonable to assume that the values $\hat{V}(j, i; L_n, B \cup A_E \cup B_E)$ are available as well; they can be readily stored as intermediate results when L_n is processed. Equally accessible (and computed only once) is $\max_{B \subseteq B_U \setminus \{n\}} \hat{V}(j, i; L_n, B \cup A_E \cup B_E)$, which can be constructed "bottom-up" when the hierarchy of lattice instances is processed, and therefore need not involve all exponentially many combinations.

With that in mind (8.95) is equivalent to (8.91), but with only $|A_U| + |B_U|$ terms instead of $2^{|A_U| + |B_U|}$. The complexity of step 3∗b (and likewise of step 3∗a) can thus be reduced dramatically, at the cost of additional storage of intermediate results. Figure 8.11 gives an idea of the savings for the case $|A_U| = |B_U| = 3$.

It is important to keep in mind, however, that the overall number of lattice instances is not reduced by this algebraic trick. Again, think of lattice

	B_U							
A_U	ccc	ccE	cEc	Ecc	cEE	EcE	EEc	EEE
ccc	max	max	max	max	max			
ccE	min							
cEc	min							
Ecc	min							
cEE								
EcE								
EEc								
EEE								

Fig. 8.11. The table shows all $2^{3+3} = 64$ early exercise combinations if $|A_U| = |B_U| = 3$. Rows represent combinations selected from A_U, and columns represent combinations picked from B_U. A lower-case "c" means continuation or no exercise, an upper-case "E" means Exercise. The formula in (8.95) examines the filled in table elements only. "max" indicates a subterm of (8.93). "min" indicates a subterm of (8.94) (the column position is slightly misleading in this case, as the correct selection from B_U might differ; it is, however, already reflected in $\hat{V}(j,i;L_m)$)

instances as vertices and associations through lookup as directed edges. The result of the algebraic transformation is to reduce the number of edges, but the set of vertices remains unaltered.

In an early stage of our research, (8.91) was replaced by (8.95). Although no rigorous tests have been made, the speedup appeared to be marginal if the number of instruments was very small, but noticeable once the number of instruments increased. Since there was no obvious drawback and the additional overhead in memory management seemed to be outweighed by the benefit in all cases, (8.95) has been used ever since.

	maintaining		collapsing
σ_{max}	time [s]	# of lattices	time [s]
0.10	0.4	1	–
0.12	1.9	6	1.0
...			
0.24	1.9	6	1.0
0.26	2.9	7	1.0
0.28	3.0	8	1.0
...			
0.40	2.7	8	1.1
0.42	4.2	10	1.0
...			
0.50	3.9	10	1.0

Fig. 8.12. Results for a portfolio of three American 30-day puts with strikes 90, 100 and 110, evaluated under a volatility band of $[0.1, \sigma_{max}]$, with σ_{max} ranging from 0.1 to 0.5 in steps of 0.02. Shown is the running time if corridors of uncertainty are maintained, together with the number of lattice instances created (of those, up to 6 lattice instances are used to monitor the corridors of uncertainty). Also shown is the running time if corridors of uncertainty are collapsed

8.3 Empirical Results

It is important to verify the value of the corridor concept in practice.

– In Sect. 8.3.1 we study the impact of the volatility range on the number of lattice instances required to solve the worst-case pricing problem for American option portfolios.
– In Sect. 8.3.2 we run a stress test on a portfolio space of 200 randomly selected American options portfolios.

In both cases the software Mtg is used. Running times are reported in seconds, but should be taken as "units of time", since they were obtained on a slow 166-MHz PC.

8.3.1 Computational Complexity

We investigate the computational complexity that arises from the positive relation between the width of the volatility band and the number of lattice instances required for the solution of the best worst-case pricing problem. In this section, scenarios and portfolios are constructed under "lab conditions", to probe certain performance characteristics while perturbing the setup as little as possible.

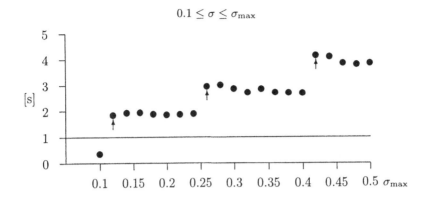

Fig. 8.13. Running times in seconds for the three 30-day American puts with strikes 90, 100 and 110, respectively, evaluated under different volatility ranges $[0.1, \sigma_{\max}]$, $0.1 \leq \sigma_{\max} \leq 0.5$. The arrows mark jumps in the number of lattice instances, due to increasing overlap of the corridors of uncertainty. The horizontal line at 1 s represents the average running time if the corridors of uncertainty are collapsed

Experiment 1: Three American Puts The portfolio consists of three 30-day American puts with strikes 90, 100 and 110, respectively. Market parameters are $S_0 = 100$ and $r = 0.03$. The size of the time step is $dt = 1/(5 \times 365)$, or five periods per day, 150 periods overall. All results are obtained with the explicit method.

Figure 8.12 gives an overview of the running time when the corridors of uncertainty are maintained (as described in Sect. 8.2.1) and collapsed (as described in Sect. 8.2.2), respectively. In the linear case ($\sigma_{\min} = \sigma_{\max}$) Mtg takes 0.4 seconds to compute the result. If corridors are maintained, the running time is stable in the intervals $0.12 \leq \sigma_{\max} \leq 0.24$ (small $\sigma_{\max}$), $0.26 \leq \sigma_{\max} \leq 0.40$ (medium $\sigma_{\max}$) and $0.42 \leq \sigma_{\max} \leq 0.5$ (large $\sigma_{\max}$), with jumps of about 1.5, 1 and 1.5 seconds preceding the intervals. Jumps correspond to the introduction of more lattice instances, as wider volatility bands lead to more overlap among corridors of uncertainty.

Figure 8.13 contains a graph of the running times. Figure 8.14 displays the location and extent of the corridors of uncertainty for the three puts schematically for the qualitatively different small, medium and large volatility ranges. For medium and large $\sigma_{\max}$, the labels in the picture indicate the non-singleton residual portfolios that are part of the early exercise combinations considered. In the medium scenario, 3 partial non-singleton portfolios need to be maintained. In the large scenario, 4 partial non-singleton portfolios need to be maintained. In addition, some singleton lattice instances may need to

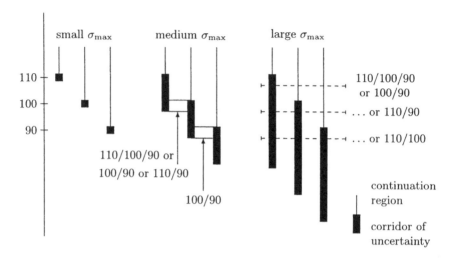

Fig. 8.14. A schematic view of the extent of the corridors of uncertainty for the three American puts mentioned in the text, under three qualitatively different volatility ranges. The vertical axis marks the value of the underlying. The labels indicate the residual non-singleton portfolios (at least 2 puts); some singleton residual portfolios are evaluated in addition to maintain the corridors of uncertainty. Instruments are identified by their strike, which is 110, 100 and 90, respectively

be maintained to feed boundary values, and some to provide the upper or lower boundary for the corridor of uncertainty.

Also compare with Fig. 8.8, which shows the actual corridor shape for some sample values of $\sigma_{\max}$. Notice that the number of combinations cited there, 4 respectively 8, refers to the maximum number of early exercise combinations that enter the minmax term at any given node instance. This number is a lower bound for the overall number of lattice instances.

The prices computed by both methods (maintaining versus collapsing) were identical. Collapsing the corridors of uncertainty, however, turns out to reduce the running time considerably and to make it independent from $\sigma_{\max}$. The speedup is approximately $3.9/1.0 \approx 4$ for $\sigma_{\max} = 0.5$.

Experiment 2: Increasing the Number of Puts In the previous experiment the size of the portfolio remained stable, while the width of the volatility range increased. In experiment 2 the number of puts in the portfolio is varied. All puts mature in 30 days and differ only by their strikes. The extremal strikes are 80 and 120, and all other strikes are equidistantly spaced between those endpoints. Thus, a portfolio of size 5 contains the strikes 80, 90, 100, 110 and 120.

8.3 Empirical Results

The number of puts varies between 2 (strikes 80 and 120 only) and 21 (10 strikes below 100, 10 strikes above 100, and 100 itself). The experiment is repeated for a linear scenario ($\sigma = 0.1$) and two nonlinear scenarios, with $\sigma_{\min} = 0.1$ and $\sigma_{\max} = 0.125$ and 0.15, respectively. All other parameters are unchanged: $S_0 = 100$, $r = 0.03$ and $dt = 1/(5 \times 365)$.

Figure 8.15 shows the running times for all three volatility scenarios and all 20 portfolio sizes. It also shows, in a superimposed step function, how the number of lattice instances grows with the number of puts, approximately following the trend of the running time. The number of lattice instances is scaled to fit into the plot, not absolute.

The significant result of this test is the validity of the concept of corridors of uncertainty: there are 2^{21} theoretical early exercise combinations for the portfolio that contains 21 puts. The running time is clearly under control.

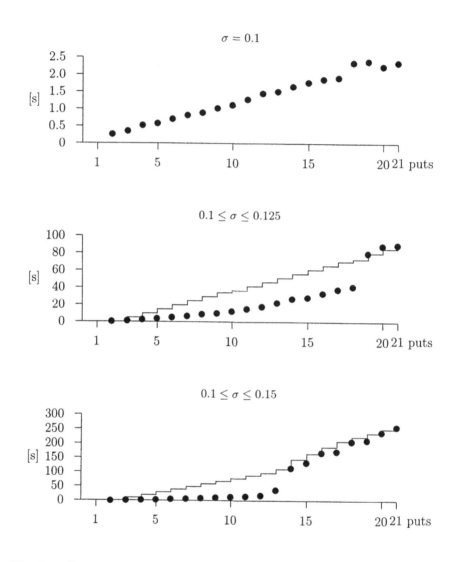

Fig. 8.15. Running times in seconds for portfolios with a varying number of 30-day American puts under three volatility scenarios. Solid disks mark running times. The step function in the bottom graphs tracks the relative (!) growth of the number of lattice instances created. In the top graph there is only one lattice instance

8.3.2 Stress Tests

The experiments in the previous section were conducted under laboratory conditions: all parameters but one remained frozen so as to test the influence of the selected parameter on the running time of Mtg. The test dimensions were the width of the volatility band and the density of strikes.

In this section we lift the ceteris paribus condition, and compute best worst-case prices for a set of portfolios with divergent characteristics. Statistical measures are then used to judge performance and accuracy, following [17].

The tests on the following pages are stress-tests. For huge volatility ranges ($\sigma_{\max} = 6\,\sigma_{\min}$) the results are less accurate. In Chapter 9 we show how this problem can be circumvented.

The Random Portfolio Space The random portfolio space consists of 200 portfolios. Each portfolio consists of 8 options with characteristics determined randomly as follows:

– With equal probability the option is a call or a put.
– For call options, the strike lies in the interval $[80, \ldots, 110]$ with probability

$$\Pr\{\text{strike} = x\} = \begin{cases} \dfrac{1}{42} & \text{if } 80 \leq x \leq 100 \\ \dfrac{1}{20} & \text{if } 101 \leq x \leq 110 \end{cases} \qquad (8.96)$$

For put options, the strike lies in the interval $[90, \ldots, 120]$ with probability

$$\Pr\{\text{strike} = x\} = \begin{cases} \dfrac{1}{20} & \text{if } 90 \leq x \leq 99 \\ \dfrac{1}{42} & \text{if } 100 \leq x \leq 120 \end{cases} \qquad (8.97)$$

Thus, in-the-money and out-of-the-money options are equally likely, but the range of possible strikes for in-the-money options is about twice as wide as for out-of-the-money options.
– The maturity is uniformly distributed in the interval $[50, \ldots, 100]$, counted in days.
– The position λ_n, $1 \leq n \leq 8$, is ± 1, ± 2 or ± 3 with equal probability.

Figure 8.16 gives a summary.

The random portfolio space is furthermore divided into two subsets:

– For the first 100 portfolios, 4 options are American. The remaining 4 options are European. We refer to this subset as the 4/4 set of portfolios.
– For the last 100 portfolios, 5 options are American. The remaining 3 options are European. We refer to this subset as the 5/3 set of portfolios.

8 Algorithms for American Options

type	probability	strike	maturity	position
call	$\frac{1}{2}$	betw. 80 and 110	betw. 50 and 100	±1, ±2 or ±3
put	$\frac{1}{2}$	betw. 90 and 120	betw. 50 and 100	±1, ±2 or ±3

Fig. 8.16. The random portfolio space. Each option has the characteristics listed in the table, randomly selected. In addition, options are either European or American

	volatility	series 1	series 2
Experiment 1	$0.1 \leq \sigma \leq 0.2$	$dt = 1/365$	$dt = 1/(2 \times 365)$
Experiment 2	$0.1 \leq \sigma \leq 0.4$	$dt = 1/365$	$dt = 1/(2 \times 365)$
Experiment 3	$0.1 \leq \sigma \leq 0.6$	$dt = 1/365$	$dt = 1/(2 \times 365)$

Fig. 8.17. The evaluation space. Each series in each experiment was performed with maintained corridors of uncertainty, and with collapsed corridors of uncertainty. The time steps are chosen to guarantee numerical stability

The random portfolio space extends the theoretical framework observed so far in two aspects:

1. Maturity dates differ; and
2. there are American and European instruments in each portfolio.

These "advanced" features are incorporated to simulate actual situations better. Both features are straightforward to add to the theoretical base. The 4 respectively 3 European options contribute to curvature through superposition of their vanilla payoff structures.

The Evaluation Space Each of the portfolios $(\mathbf{X}_1^{44}, \lambda_1^{44})$, ..., $(\mathbf{X}_{100}^{44}, \lambda_{100}^{44})$ in the 4/4 set, and $(\mathbf{X}_1^{53}, \lambda_1^{53})$, ..., $(\mathbf{X}_{100}^{53}, \lambda_{100}^{53})$ in the 5/3 set was evaluated several times under varying conditions. Always, however, $\sigma_{\min} = 0.1$. The market parameters are $S_0 = 100$, interest rate $r = 0.05$ and dividend rate $q = 0.03$ throughout. (A dividend rate is introduced since the portfolios contain American calls which are not exercised early if $q = 0$.)

All portfolios as well as their negative versions ($-\lambda$ instead of λ) were evaluated in three experiments under the three volatility scenarios $\sigma_{\max} = 0.2$, 0.4 and 0.6, respectively. In each experiment, two series of evaluations were performed with maintained corridors of uncertainty, and two series of evaluations were performed with collapsed corridors of uncertainty. The time step in series 1 was $dt = 1/365$, under both methods; in series 2 it was $dt = 1/(2 \times 365)$. Altogether this experiment required $2 \times 2 \times 3 \times (2 \times 100 + 2 \times 100) = 4800$ evaluations.

Figure 8.17 summarizes all three experiment specifications.

8.3 Empirical Results 117

experiment	σ_{max}	time step	4/4 set mean	4/4 set sdev	5/3 set mean	5/3 set sdev
experiment 1, series 1	0.2	1/365	1.5	1.0	3.4	2.8
experiment 2, series 1	0.4	1/365	3.1	1.1	9.5	4.2
experiment 3, series 1	0.6	1/365	3.5	1.1	10.7	4.1

Fig. 8.18. The running time in seconds if corridors of uncertainty are maintained, broken down for the 4/4 set and the 5/3 set of portfolios. Each entry represents $2 \times 100 = 200$ evaluations, as in (original + negative) × portfolios

Maintaining the Corridors of Uncertainty We give absolute results for the exact approach where corridors of uncertainty are maintained as described in Sect. 8.2.1. In a later paragraph, the benefit and drawback of collapsing the corridors is analyzed relative to the absolute values given here.

Figure 8.18 presents the mean and standard deviation of the running time if corridors of uncertainty are maintained. Only series 1 is analyzed; no data is available for series 2.

Figure 8.19 presents data on convergence with respect to the time step. The results obtained in series 1 and 2 are matched and compared pair-wise. Shown are the first two central moments, in percentage, of

$$\frac{\hat{V}_2(0,0;L_n^{44}) - \hat{V}_1(0,0;L_n^{44})}{\hat{V}_1(0,0;L_n^{44})} \tag{8.98}$$

for $1 \leq n \leq 200$, corresponding to 2×100 portfolios (including the negative λ's) in the 4/4 set, and the first two central moments, in percentage, of

$$\frac{\hat{V}_2(0,0;L_n^{53}) - \hat{V}_1(0,0;L_n^{53})}{\hat{V}_1(0,0;L_n^{53})} \tag{8.99}$$

for $1 \leq n \leq 200$, corresponding to 2×100 portfolios in the 5/3 set. Here, $\hat{V}_l$ is the best worst-case price observed in series l. L_n^{44} is the lattice instance with signature $(\mathbf{X}_n^{44}, \lambda_n^{44})$ if $n \leq 100$, and with signature $(\mathbf{X}_{n-100}^{44}, -\lambda_{n-100}^{44})$ if $n \geq 101$. L_n^{53} is interpreted in an analogue fashion.

Convergence is better for narrow volatility bands. For $\sigma_{max} = 0.2$ we may expect stability in the first two leading digits, and thus recommend $dt = 1/(2 \times 365)$ as adequate. On the other hand, there is considerable variance if $\sigma_{max} \geq 0.4$. This suggests that for wide volatility ranges dt needs to be further reduced to achieve sufficient numerical stability.

Collapsing the Corridors of Uncertainty: Speed After establishing a base for comparison, we examine the benefit of collapsing corridors of uncertainty. Let m_n^{44} be the running time if corridors of uncertainty are maintained for portfolio n (where $1 \leq n \leq 200$, and portfolios are counted as described

	$\sigma_{\max}$	time steps	4/4 set mean	sdev	5/3 set mean	sdev
experiment 1	0.2	from $1/365$ to $1/(2\times 365)$	0.1	0.6	0.2	1.6
experiment 2	0.4	from $1/365$ to $1/(2\times 365)$	0.3	6.0	-0.8	7.5
experiment 3	0.6	from $1/365$ to $1/(2\times 365)$	0.0	11.2	-0.2	2.8

Fig. 8.19. The relative discrepency in percentage for each matched evaluation within series 1 and 2, respectively, in each experiment, broken down by volatility band and portfolio set. The number of time steps is doubled between series 1 and 2

	$\sigma_{\max}$	time step	4/4 set mean	sdev	5/3 set mean	sdev
experiment 1, series 1	0.2	$1/365$	63.5	10.8	55.5	11.7
experiment 2, series 1	0.4	$1/365$	66.1	13.6	55.1	18.2
experiment 3, series 1	0.6	$1/365$	62.5	16.3	52.6	18.2

Fig. 8.20. Mean and standard deviation in percentage of the relative running time if corridors of uncertainty are collapsed, broken down by volatility band and portfolio subset. The last row is the average over all previous rows. The inverse of the mean would be the average speedup factor

above) in the 4/4 set, and let c_n^{44} be the running time if corridors of uncertainty are collapsed. m_n^{53} and c_n^{53} are interpreted accordingly. Figure 8.20 shows mean and standard deviation for the quantities

$$\frac{c_n^{44}}{m_n^{44}} \quad \text{and} \quad \frac{c_n^{53}}{m_n^{53}} \qquad (8.100)$$

in percentage, for $1 \leq n \leq 200$, broken down by experiment and series, as well as aggregated over all experiments. Figure 8.21 shows the same data pictorially.

The relative benefit is remarkably uniform for different volatility bands, although the benefit decreases slightly for very high $\sigma_{\max}$. The standard deviation is under 20% throughout. Relative speed increases if portfolios contain more American instruments.

Collapsing the Corridors of Uncertainty: Faithfulness Collapsing the corridor of uncertainty may lead to false results. The faithfulness of the heuristic measures the gravity of this defect. Let L_n^{44} and L_n^{53} denote lattice instances for portfolios $1 \leq n \leq 200$ in the 4/4 and the 5/3 set, respectively, as defined earlier. Let the benchmark result $\hat{M}(0,0;L_n^{44}) = \hat{V}(0,0;L_n^{44})$ be the best worst-case price on lattice instance L_n^{44} if corridors of uncertainty are maintained, and define $\hat{M}(0,0;L_n^{53})$ accordingly. Let $\hat{C}(0,0;L_n^{44})$ be the best worst-case price if corridors are collapsed, and define $\hat{C}(0,0;L_n^{53})$ accordingly.

8.3 Empirical Results 119

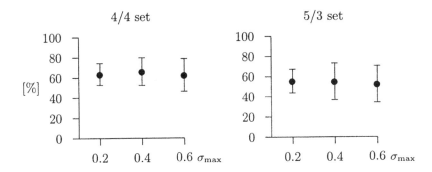

Fig. 8.21. Mean ± one standard deviation of the relative running time in percentage when corridors of uncertainty are collapsed, compared to the running time for the benchmark result. The data is the same as in Fig. 8.20

	σ_{max}	time step	4/4 set			5/3 set		
			mean	sdev	good	mean	sdev	good
experiment 1, series 2	0.2	$1/(2 \times 365)$	0.00	0.32	99.0	−0.29	4.06	98.0
experiment 2, series 2	0.4	$1/(2 \times 365)$	0.85	7.94	93.0	0.07	2.15	94.0
experiment 3, series 2	0.6	$1/(2 \times 365)$	−0.12	9.22	90.5	−0.22	2.44	88.5

Fig. 8.22. Mean and standard deviation in percentage of the relative deviation from the benchmark result if corridors of uncertainty are collapsed. The column labeled "good" shows the frequency with which the benchmark result is reproduced exactly

$\hat{C}(0,0;L_n^{44})$ and $\hat{C}(0,0;L_n^{53})$ may differ from $\hat{V}(0,0;L_n^{44})$ and $\hat{V}(0,0;L_n^{53})$. The faithfulness of the heuristic is reflected in the relative deviation from the benchmark result:

$$\frac{\hat{C}(0,0;L_n^{44}) - \hat{M}(0,0;L_n^{44})}{\hat{M}(0,0;L_n^{44})} \tag{8.101}$$

and

$$\frac{\hat{C}(0,0;L_n^{53}) - \hat{M}(0,0;L_n^{53})}{\hat{M}(0,0;L_n^{53})} \tag{8.102}$$

Values close to 0 indicate high faithfulness. Large absolute values indicate low faithfulness. Mean and standard deviation in percentage of (8.101) and (8.102) are shown in Fig. 8.22 for series 2. Also shown is the frequency in percentage with which the approximated result deviates no more than 1% from the benchmark result.

120 8 Algorithms for American Options

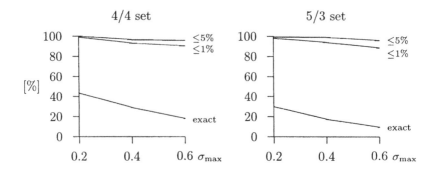

Fig. 8.23. Frequency in percentage with which the relative error stays within 0% (exact match), 1% and 5% of the benchmark result if corridors of uncertainty are collapsed, drawn against $\sigma_{\max}$. Data is based on series 2 evaluations

$\sigma_{\max}$	set	deviation [%]	n	exercise	type	maturity	strike	λ_n
0.4	4/4	107.4	1	European	put	100	92	−3
→ 0.6	4/4	88.1	2	European	put	68	97	2
0.6	4/4	−85.3	3	European	call	66	93	−1
0.2	5/3	−57.4	4	European	put	61	98	1
			5	American	put	97	113	−3
			6	American	put	93	114	2
			7	American	call	68	102	1
			8	American	put	57	93	−2

Fig. 8.24. Four cases in experiment 1, series 2 (time step $dt = 1/(2 \times 365)$), in which the relative deviation from the benchmark result exceeds 50%, and the composition of one of the outlier portfolios (marked with "→")

Figure 8.23 plots the frequency of exactly matching results and of results that deviate no more than 1 or 5% (the frequency of exact matches is not shown in Fig. 8.22). The frequency of "good" results drops consistently as the volatility band gets wider, and slightly if the portfolio contains more American options. Although the heuristic reproduces the benchmark result less than half the time, the frequency at which a 1% relative error bound is achieved is above or close to 90% throughout.

Collapsing the Corridors of Uncertainty: Outliers There are 4 cases in the series 2 evaluations in which the absolute deviation from the benchmark result exceeds 50%. The amount by which these cases deviate is shown in Fig. 8.24, together with the composition of one of the outlier portfolios.

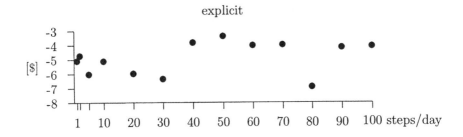

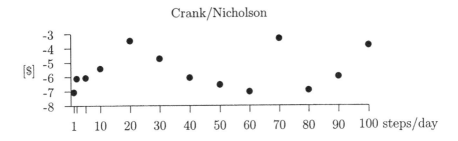

Fig. 8.25. High oscillation of the best worst-case value of the portfolio marked in Fig. 8.24 even if the corridors of uncertainty are maintained. There is no qualitative difference between the explicit and Crank-Nicholson scheme. The number of steps per day d is plotted on the x-axis; the corresponding time step would be $dt = 1/(d \times 365)$

The sequence of best worst-case prices for the marked portfolio shows considerable oscillation as the time step decreases, even if the corridors of uncertainty are maintained. Fig. 8.25 shows the values, plotted against the number d of steps per day for both the explicit and mixed explicit/implicit scheme (this is the only case in this chapter on American options where Crank-Nicholson combined with iterative refinement is used). 20 lattice instances, the largest possible number, are required to solve the best worst-case pricing problem. The inverse portfolio ($-\lambda_n$ instead of λ_n for $1 \leq n \leq 8$) converges convincingly: the value varies around -36.19, with noise in the fourth digit (values are not shown here). There is no obvious sign that helps to explain what makes the portfolio structurally unusual enough to lead to such instability.

Further comparative convergence analysis with 1, 2, 5 and 20 time steps per day, for all portfolios in the 4/4 subset under the scenario $\sigma_{\max} = 0.6$, shows that there is no correlation between poor numerical convergence and a large deviation from the benchmark result in the series 2 data. The Spearman and Kendall rank coefficients for the association between the absolute relative change of the benchmark result when switching from 1 to 20 time steps per day, and the maximum absolute deviation from the benchmark in the series 2 data are 0.09 and 0.06, respectively. Rank coefficients measure linear and nonlinear monotonic relationships. A value close to zero means there is no such relationship. The linear correlation coefficient is 0.17.

The quality of the result achieved under the heuristic seems therefore unpredictable.

Summary The benefits of collapsing the corridor of uncertainty seem worth the loss of faithfulness if the volatility band is narrow, for then the benchmark results are reproduced to a sufficiently high degree. For $0.1 \leq \sigma \leq 0.2$, for instance, the mean error is zero and the standard deviation of the error is 0.32%, for 4 American options in the portfolio (Fig. 8.22). This is equivalent to 2 matching digits.

The situation becomes less clear as $\sigma_{\max}$ increases. Whether the gain in speed of about 40% is worth the increased chance of missing the best worst-case price by a large amount must be decided case by case. As shown in Fig. 8.23, the 1% or 2 leading digit-threshold is still reached about 90% of the time.

It should be noted that the volatility bands used in the mass test are very wide and remain valid over the entire lifetime of the portfolio. In a more realistic setting, the range of uncertainty would be narrower or restricted in time. The next chapter explores volatility scenarios in this direction.

Numerical accuracy at timesteps in the tested range is satisfactory for narrow bands (Fig. 8.19). For wider bands, smaller time steps than those tested should be used in production mode. The use of the more accurate mixed explicit/implicit finite difference scheme would improve the convergence behavior further.

8.4 American Options and Calibration

It is in principle possible to apply the ideas of Sect. 4.2.3 on calibration to portfolios of American options. The calibrated volatility $\hat{\sigma}$ would be path-dependent and not easily convertible into a two-dimensional surface. On the other hand, the volatility surface, being the goal of calibration in the first place, should have a format in which subsequent pricing is straightforward. Calibration to American options seems therefore not a worthwhile task.

Optimizing a position in order to find the optimal hedge portfolio under worst-case assumptions, on the other hand, would still be feasible (see Sect. 4.2.2).

9 Exotic Volatility Scenarios

In Chapters 7 and 8, algorithms have been discussed that compute (best) worst-case prices under uncertain volatility scenarios in which $\sigma(S_t, t)$ and $\sigma(S_u, u)$ are independent for $t \neq u$. In this chapter we extend the notion of uncertain volatility scenarios to include evolutions of the spot volatility that depend on its past history.

The path-dependent character of σ is expressed in by-conditions in the candidate set $\mathcal{C}$. σ no longer depends merely on S_t and t, but on the path ω in the probability space. Replicating the terminology for instruments, we call such volatility scenarios exotic volatility scenarios, as opposed to "conventional" volatility scenarios. In particular, we examine scenarios where the spot volatility can undergo one or several volatility shocks of limited duration.

9.1 Volatility Shocks for Portfolios of Vanilla Options

Volatility shock scenarios are based on the assumption that the spot volatility does not deviate from an estimated prior volatility except possibly when expected or unexpected economic events upset the market for a limited period of time. Such events may be announcements, mergers, court rulings, natural disasters, devaluations, or others. These events have the properties that

- they are difficult to quantize; and, more importantly,
- they cannot be forecasted to happen on a specific day in the future.

We use the worst-case approach for the quantization problem, and multi-lattice dynamic programming for the forecasting problem.

Definition 21 (Prior and shock volatility). *Assume we are given volatility values $0 < \sigma_{\min} \leq \sigma_0 \leq \sigma_{\max}$. Then σ_0 is called the* prior volatility *and expresses the subjective belief of the agent about the true model volatility. $\sigma_{\min}$ and $\sigma_{\max}$ are lower and upper bounds which the spot volatility can attain during periods of upheaval. They are called the* shock volatility *bounds.*

For simplicity, Def. 21 introduces constant volatility parameters. The concepts in this chapter can easily be extended to cover time and/or space dependent prior and shock volatilities. (Recall that this does not mean that the worst-case volatility is also constant!)

Definition 22 (Volatility shock scenario). *Assume prior and shock volatility $0 < \sigma_{\min} \leq \sigma_0 \leq \sigma_{\max}$ are given. A volatility shock scenario is characterized by its duration $d \geq 1$, its periodicity $p \geq 1$ and its frequency $f \geq 1$. The units of d and p are days; f is a dimensionless number. All values are integers.*

The interpretation is as follows: on any realized path ω the spot volatility will be σ_0, except for f non-overlapping periods of length d days each, during which the spot volatility may fluctuate freely within $\sigma_{\min}$ and $\sigma_{\max}$. Here, "non-overlapping" refers to the interior of each period; they may touch at their endpoints. In addition, each of these f shock periods must start on a day whose day count number is a multiple of p, where days are counted from 0.

The class of volatilities that fulfill this description is denoted by $\mathcal{D}$.

The function of p is to reduce the computational overhead and the size of the lattice. We will see below that the compute time is proportional to d/p. p may also be used to time shock periods, but to support this aspect fully a more powerful notion of periodicity may be necessary. Although in most cases $p \leq d$, we explicitly allow the case $p > d$. The f shock periods are located between time 0 and time N. In the following, we assume $N \geq d + (f-1)\max(p,d)$ for convenience. In other words, the portfolios under investigation last long enough to fall under the influence of at least f shock periods.

Examples of volatility shock scenarios are:

- The prior volatility is $\sigma_0 = 0.15$. However, there will be a 7-day period during which the volatility may oscillate between 0.15 and 1.0. This period, caused by a merger announcement expected in the near future, can start on any day. Thus, $\sigma_0 = \sigma_{\min} = 0.15$, $\sigma_{\max} = 1.0$, $d = 7$, $p = 1$, $f = 1$.
- The central bank of country XYZ meets once a week. It is expected that an important economic decision will be made in one of its future meetings, though it is not known in which one. Heavy trading on the day following the announcement is anticipated. In this case, $\sigma_0 = \sigma_{\min} = 0.15$, $\sigma_{\max} = 1.0$, $d = 1$, $p = 7$, $f = 1$ may be a realistic volatility shock scenario.

The crucial property of volatility shock scenarios is that they leave open when the shock periods occur. If the timing of events is known, a time-dependent conventional uncertain volatility scenario works adequately. It is the additional dimension of uncertainty of timing which opens the door to worst-case considerations.

The other quantitative difference between conventional and volatility shock scenarios is the width of the volatility band: while conventional scenarios may allocate a 0.1–0.2 volatility band, for instance, volatility shock scenarios provide for volatility spikes of much larger amplitude. Wide bands in the conventional scenario suffer from two flaws: a) they lead to wide price bands, and b) they do not reflect the isolated nature of events which influence market behavior. Volatility shock scenarios alleviate both drawbacks.

9.1.1 Worst-Case Volatility Shocks

Under the worst-case paradigm volatility shock periods are located such that the resulting worst-case price is maximized. The market is regarded as adversary that triggers events perturbing the prior volatility at the most adverse moment.

The objective of worst-case pricing under a conventional volatility scenario has been formulated in Sect. 4.2:

Given a portfolio $\mathbf{X}$ and a position $\lambda \in \mathbb{R}^k$ in $\mathbf{X}$, which $\sigma \in \mathcal{C}$ maximizes today's value of $(\mathbf{X}, \lambda)$?

The extension to volatility shock scenarios is straightforward and goes as follows.

Given a portfolio $\mathbf{X}$ and a position $\lambda \in \mathbb{R}^k$ in $\mathbf{X}$. Given furthermore prior and shock volatilities $\sigma_{\min} \leq \sigma_0 \leq \sigma_{\max}$ and shock scenario attributes d, p and f. Which $\sigma \in \mathcal{D}$ maximizes today's value of $(\mathbf{X}, \lambda)$?

$\mathcal{D}$ has been defined in Def. 22 as the class of volatilities σ that satisfy

$$\sigma_{\min} \leq \sigma(\omega, t) \leq \sigma_{\max} \tag{9.1}$$

during shock periods and

$$\sigma(\omega, t) = \sigma_0 \tag{9.2}$$

during silent periods. We assume that $\mathbf{X}$ contains only vanilla options, all maturing at time t_N.

Multi-Lattice Dynamic Programming Revisited The worst-case volatility-shock pricing problem can be solved with multi-lattice dynamic programming. The number of lattice instances depends on the volatility shock scenario and can be known beforehand. Each lattice instance carries $(\mathbf{X}, \lambda)$, but solves PDE (4.9) with a different, non-path-dependent (!) volatility coefficient. Transferring data between lattice instances works much like in the American case: local decisions are made with regard to the "shock front", i.e. the optimal (that is, worst) time of entering a shock period. The shock front is the analogue of the early exercise boundary.

Figure 9.1 gives an example. Lattice instance L is the top-level lattice instance yielding the final result $\hat{V}(0,0;L)$. Paths 1, 2 and 3 originating at s_0 and hitting the shock front at time t_1 (paths 1 and 2) respectively at some later, unspecified time (path 3) are traced. After hitting the shock front, paths 1 and 2 continue on lattice instance L'. L' differs from L in that it prices with the conventional uncertain volatility scenario

$$\begin{aligned} \sigma(S_t, t) &= \sigma_0 & (t < t_1 \text{ or } t > t_2) \\ \sigma_{\min} \leq \sigma(S_t, t) &\leq \sigma_{\max} & (t_1 \leq t \leq t_2) \end{aligned} \tag{9.3}$$

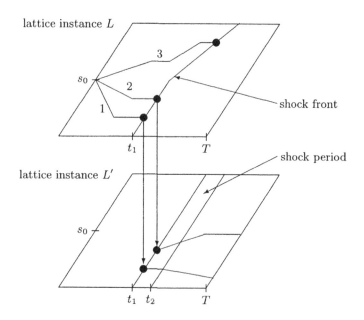

Fig. 9.1. Paths 1 and 2 hit the shock front at time t_1 and switch to lattice instance L', which solves a conventional worst-case pricing problem with time-dependent $\sigma_{\min}$ and $\sigma_{\max}$ (i.e., $\sigma_{\min} = \sigma_{\max} = \sigma_0$ for $t < t_1$ and $t > t_2$). Path 3 hits the shock front at a later time and continues on a lattice instance with a different conventional worst-case volatility scenario

with a fixed period of volatility oscillation between times t_1 and t_2. L, on the other hand, prices with σ_0 between $t = 0$ and the shock front, whose location is determined with the dynamic programming method.

Path 3 does not hit the shock front at time t_1 and therefore does not continue on L', but on another lattice instance whose shock period is located suitably. Notice that while the shock front in L is uneven, the shock period in L' itself starts uniformly at time t_1 and ends uniformly at time t_2.

The example seems to suggest that there must be a lattice instance for every possible location of the shock period. This is not so; lattice instances can be reset and reused in the rollback scheme as soon as a shock period is finished. A combination of high-level handling of lattice instances and conventional worst-case pricing is powerful enough to solve the worst-case pricing problem under volatility shock scenarios.

Definition 23 (Extended lattice signature). *Given* $(\mathbf{X}, \lambda)$, *duration* d, *periodicity* p *and frequency* f. *The extended signature of a lattice instance* L *for the so-specified volatility shock scenario is a quintuple* $(\mathbf{X}, \lambda, \tau, \xi, \delta)$, *where* $\tau \in \{\text{conventional}, \text{consolidate}\}$ *is the type,* $0 \leq \xi \leq f$ *is the level, and*

9.1 Volatility Shocks for Portfolios of Vanilla Options

$0 \leq \delta \leq \lceil d/p \rceil$ is the offset of the lattice instance. The offset is undefined if $\tau =$ consolidate.

If **X** contains only vanilla options, all lattice instances carry the same portfolio $(\mathbf{X}, \lambda)$. In that case, we ommit **X** and λ and write (τ, ξ, δ).

Consolidating lattice instances use subordinate conventional lattice instances to locate the shock front. If the duration exceeds the periodicity, potential shock periods may overlap, and up to $\lceil d/p \rceil$ conventional lattice instances need to be maintained to feed a single consolidating lattice instance. Consolidating and associated conventional lattice instances are grouped in levels. Levels are ordered, for conventional lattice instances, in turn, fetch their boundary data from lower level consolidating lattice instances. Thus, L in Fig. 9.1 is consolidating while L' is conventional.

Level 0 is unique in that it does not contain any conventional lattice instances. The consolidating lattice instance of level 0 prices $(\mathbf{X}, \lambda)$ by definition with σ_0. On level 0, pricing becomes linear.

Figure 9.2 explains these concepts for $d = 4$, $p = 2$ and $f = 1$. Shock periods are possible between days 0–4, 2–6, 4–8, 6–10 and 8–10 (the last one being cut off at day 10). The main lattice instance L_1 imports worst-case prices on days 0, 2, 4, 6 and 8 from conventional lattice instances L_1^0 and L_1^1, depending on the offset. After maximizing locally just like it is done for American options, the resulting value is rolled back 2 days with linear volatility $\sigma = \sigma_0$. Then data is imported from L_1^0 or L_1^1 and compared again. The shock front is implicitly given by the outcome of the local maximization operations and continuously re-adjusted.

The conventional lattice instances L_1^0 and L_1^1 are reused several times. After worst-case prices have been transferred to L_1 on days 0, 2, 4, 6 and 8, the lattice instances are reset with current linear prices, copied from L_0. Here and in the subsequent paragraphs, "current" refers to the loop variable i which iterates through time slices $N, \ldots, 0$ (i is part of the input in the algorithm in Fig. 5.5). The function of L_1, L_1^0, L_1^1 and L_0 can be summarized, bottom-up, as follows:

- L_0 is the lattice instance at the lowest level and is used to price $(\mathbf{X}, \lambda)$ at the prior volatility $\sigma = \sigma_0$.
- L_1^0 is used to price $(\mathbf{X}, \lambda)$ under the conventional worst-case volatility scenario with a volatility band $\sigma_{\min} \leq \sigma \leq \sigma_{\max}$ during the current shock period $[2lp, 2lp + 4]$, $l \geq 0$ chosen suitably, and $\sigma = \sigma_0$ during the tail period $[2lp + 4, 10]$. The offset of L_1^1 is $\delta = 0$. As the tail period becomes longer and a volatility shock date is crossed, L_1^0 is reset with data from L_0.
- L_1^1 is used to price $(\mathbf{X}, \lambda)$ under the conventional worst-case volatility scenario with a volatility band $\sigma_{\min} \leq \sigma \leq \sigma_{\max}$ during the current shock period $[(2l + 1)p, (2l + 1)p + 4]$, $l \geq 0$ chosen suitably, and $\sigma = \sigma_0$ during the tail period $[(2l+1)p+4, 10]$. The offset of L_1^0 is $\delta = 1$, corresponding to a shift of $\delta p = 2$ days of shock periods. As the tail period becomes longer and a volatility shock date is crossed, L_1^1 is reset with data from L_0.

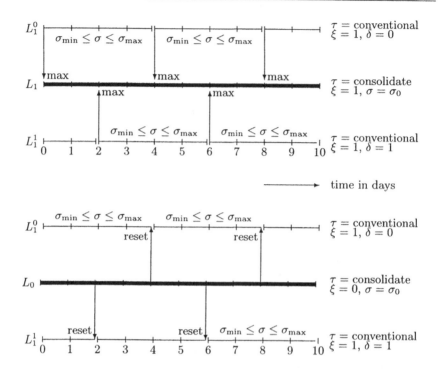

Fig. 9.2. Four lattice instances L_1, L_1^0, L_1^1 and L_0 are needed to solve a volatility shock scenario with shock duration $d = 4$, periodicity $p = 2$ and frequency $f = 1$. L_1^0 is responsible for the shock periods $[2lp, 2lp + 4]$, and L_1^1 is responsible for the shock periods $[(2l+1)p, (2l+1)p + 4]$, where $l \geq 0$. After the worst-case price for a shock period has been incorporated into the main lattice instance L_1 through local maximization (top picture), the associated conventional lattice instance is reset with the current linear price obtained with the prior volatility σ_0 (bottom picture)

- L_1 holds prices for $(\mathbf{X}, \lambda)$ which, during rollback, represent the expected payoff under the assumption that the volatility shock period has occurred sometime between the current time slice and day 10. As the rollback proceeds from day 10 to day 0, this assumption is periodically verified by checking whether the price for $(\mathbf{X}, \lambda)$ increases if the volatility shock period starts at the current time slice.

To decrease the periodicity p from 2 to 1 requires two additional conventional lattice instances for shock periods with offsets 1 and 3, respectively. The resulting cycle of periods is $[(4l + o)p, (4l + o)p + 4]$, $0 \leq o \leq 3$ and $l \geq 0$. To increase p from 2 to 4, on the other hand, makes L_1^1 superfluous, and three lattice instances overall suffice. Also note that the days on which shock periods start and end must be matched by the lattice: if d and/or p are small,

9.1 Volatility Shocks for Portfolios of Vanilla Options

the discretization becomes necessarily denser. $1/p$ is proportional to the time complexity of the pricing problem. Fine-tuning of both d and p can lead to a significant gain in response time.

If the shock frequency f is increased from 1 to 2, a new level $\xi = 2$ needs to be added. L_2 becomes the main lattice, and $\hat{V}(0,0;L_2)$ the overall result. L_2 is interpreted as follows:

- L_2 holds prices for $(\mathbf{X}, \lambda)$ which, during rollback, represent the expected payoff under the assumption that up to *two* volatility shock periods occur sometime between the current time slice and day 10.

The new conventional lattice instances L_2^0 and L_2^1 with signatures ($\tau =$ conventional, $\xi = 2, \delta = 0$) and ($\tau =$ conventional, $\xi = 2, \delta = 1$), respectively, are reset with data from L_1 when shock dates are crossed. They are interpreted as follows:

- L_2^0 prices $(\mathbf{X}, \lambda)$ under the conventional worst-case volatility scenario with a volatility band $\sigma_{\min} \leq \sigma \leq \sigma_{\max}$ during the current shock period $[2lp, 2lp + 4]$, $l \geq 0$ chosen suitably, and under the assumption that an additional shock period occurs during the tail period $[2lp + 4, 10]$.
- L_2^1 prices $(\mathbf{X}, \lambda)$ under the conventional worst-case volatility scenario with a volatility band $\sigma_{\min} \leq \sigma \leq \sigma_{\max}$ during the current shock period $[(2l + 1)p, (2l + 1)p + 4]$, $l \geq 0$ chosen suitably, and under the assumption that an additional shock period occurs during the tail period $[(2l + 1)p + 4, 10]$.

Care has to be taken that L_2^0 and L_2^1 are reset with data from L_1 only after L_1 has been processed: the data must reflect the result of the local maximization at L_1 on the shock date.

Figure 9.3 gives a schematic overview over the hierarchy of lattice instances for general f. Each consolidating lattice instance L_n, $0 \leq n \leq f$, carries the full solution of a worst-case volatility-shock pricing problem with frequency $f' = n$.

Algorithms In the following we assume a discretization that coincides with day boundaries: $t_i = i$ for $0 \leq i \leq N$. Depending on the duration and periodicity of the shock volatility scenario, this convention may be relaxed in an actual implementation.

The algorithm in Fig. 9.4 computes the required number of conventional lattice instances, and creates all lattice instances. The following lemma shows that the algorithm creates the necessary number of lattice instances, and uses them optimally.

Lemma 1 (Lattice instance creation). *Given a volatility shock scenario with duration d, periodicity p and frequency f. For any given level n, $1 \leq n \leq f$, the algorithm in Fig. 9.4 facilitates an assignment of shock periods to conventional lattice instances such that no two overlapping shock periods are assigned to the same lattice instance. (Touching at the endpoints is allowed.)*

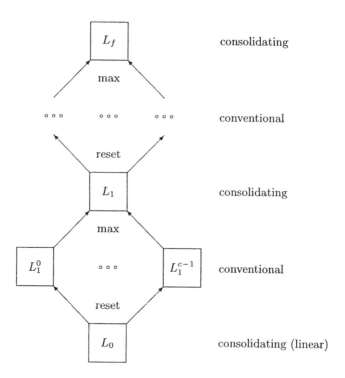

Fig. 9.3. The hierarchy of lattice instances for general f. Arrows represent the dataflow. c is the number of conventional lattice instances per level. The "max" and "reset" labels correspond to the "max" and "reset" operations in Fig. 9.2

Proof. Any shock period can be written $[lp, lp + d]$, $l \geq 0$. The quantity $c = \lceil d/p \rceil$ defined in step 1 of the algorithm is the smallest number such that $cp \geq d$, for

$$cp = \lceil d/p \rceil p \geq (d/p)p = d \qquad (9.4)$$

on one side, and

$$(c-1)p = (\lceil d/p \rceil - 1)p < ((d/p + 1) - 1)p = d \qquad (9.5)$$

in the other direction.

Now fix a level n. Consider the first c shock periods $[lp, lp + d]$, $0 \leq l \leq c - 1$. All c shock periods overlap, for their start dates $0, p, 2p, \ldots, (c-1)p$ all lie within the first period $[0, d]$, as shown in (9.5). Thus, at least c lattice instances are required to fulfill the condition that shock periods assigned to the same lattice instance do not overlap. We assign each of the c shock periods to a separate lattice instance.

9.1 Volatility Shocks for Portfolios of Vanilla Options 131

Input: Duration d, periodicity p, frequency f
Output: A set of lattice instances

1. Set $c = \lceil d/p \rceil$. c is the number of conventional lattice instances per level
2. Create lattice instance L_0 with signature

   $(\tau = \text{consolidate}, \xi = 0, \delta = \text{undefined})$

3. Repeat for $n = 1, \ldots, f$:
   a) Create lattice instance L_n with signature

   $(\tau = \text{consolidate}, \xi = n, \delta = \text{undefined})$

   b) Repeat for $m = 0, \ldots, c-1$:
      i. Create lattice instance L_n^m with signature

      $(\tau = \text{conventional}, \xi = n, \delta = m)$

Fig. 9.4. The algorithm to create all required lattice instances for a given volatility shock scenario

Let $L_n^0, \ldots, L_n^{c-1}$ be the lattice instances created. The next shock period that needs assignment is $[cp, cp + d]$. Since $cp \geq d$, assignment of this shock period to L_n^0 does not violate the no-overlap condition (although the periods may touch at their endpoints). It is easy to see how the round-robin assignment proceeds.

In summary, if day i is divisible by p, i.e. is a day on which a shock period may start, then the lattice instance to which this shock period is assigned within any given level is $m = i/p \mod c$. □

The worst-case volatility-shock pricing problem is solved in two phases. In phase 1, values are rolled back in whatever scheme has been selected (explicit or mixed explicit/implicit). In addition, local maximization is performed for consolidating lattice instances if the processed time slice falls on day on which a shock period starts. During phase 1, lower level lattice instances are processed first, and conventional lattice instances are processed before the consolidating lattice instance within the same level. This rule is an extension of the external consistency rule proposed in Sect. 5.1. Figure 9.5 shows the algorithm.

Phase 2 is dedicated to resetting the conventional lattice instances, depending on whether their offsets δ and the round-robin index $i/p \mod c$ of the shock start-date match. No particular order needs to be observed in phase 2. The data collected from consolidating lattice instances has been prepared in phase 1. The algorithm is shown in Fig. 9.6.

Instead of formalizing the notion of volatility shocks any further, we use the algorithms in Figs. 9.5 and 9.6 to define the worst-case volatility-shock

Input: Lattice instance L with signature (τ, ξ, δ), time i
Output: $\hat{V}(j, i; L)$ for $D \leq j \leq U$

1. If $\tau =$ conventional:
   a) Use the algorithm in Fig. 5.5 to get worst-case values $\hat{V}(j, i; L)$ for $D \leq j \leq U$
2. If $\tau =$ consolidate:
   a) Apply the algorithm in Fig. 5.5 with $\sigma_{\min}$ and $\sigma_{\max}$ set to σ_0 (essentially, a linearized version of the algorithm) to get initial $\hat{V}(j, i; L)$ for $D \leq j \leq U$
   b) If $\xi > 0$ and i is divisible by the periodicity p:
      i. With $c = \lceil d/p \rceil$ and $m = i/p \mod c$, repeat for $D \leq j \leq U$:

      $$\hat{V}(j, i; L) := \max\left[\hat{V}(j, i; L), \hat{V}(j, i; L_\xi^m)\right]$$

      where we can be sure that conventional L_ξ^m has already been processed earlier in phase 1; adjust the gradient of $\hat{V}(j, i; L)$ accordingly

Fig. 9.5. Phase-1 algorithm, applied to all lattice instances in the order L_0, L_1^0, ..., L_1^{c-1}, L_1, ..., L_f. Note that $\hat{V}(j, i; L)$ is treated as a variable which can be modified

---

Input: Lattice instance L with signature (τ, ξ, δ), time i
Output: Adjusted $\hat{V}(j, i; L)$ for $D \leq j \leq U$

1. If $\tau =$ conventional:
   a) If i is divisible by the periodicity p:
      i. With $c = \lceil d/p \rceil$, set $m = i/p \mod c$
      ii. If $m = \delta$ repeat for $D \leq j \leq U$:

      $$\hat{V}(j, i; L) := \hat{V}(j, i; L_{\xi - 1})$$

      where consolidating $L_{\xi-1}$ has been processed in phase 1; reset the gradient accordingly

Fig. 9.6. Phase-2 algorithm, applied to all conventional lattice instances

---

price of a portfolio. The consistency of the algorithms is clear by Lemma 1 and inspection. They are a straightforward extension of the basic concepts developed in Chapter 5.

Definition 24 (Worst-case volatility-shock price). *Given a volatility shock scenario with duration d, periodicity p and frequency f, together with prior volatility σ_0 and shock volatility bounds $\sigma_{\min}$ and $\sigma_{\max}$. The value obtained for a portfolio $(\mathbf{X}, \lambda)$ by running the algorithms in Figs. 9.5 and 9.6, embedded in a multi-lattice dynamic-programming framework as discussed in Chapter 5, is called the* worst-case volatility-shock price *of $(\mathbf{X}, \lambda)$.*

9.1 Volatility Shocks for Portfolios of Vanilla Options

In particular, the sub-additivity of the worst-case price asserted in Fact 6 (and repeated later, for American options, in Prop. 5) is maintained through the application of the maximum operator in step 2(b)i in Fig. 9.5.

Numerical Issues Volatility shock scenarios encourage short volatility spikes with large amplitude. Since these spikes can be located anywhere on the lattice, $\sigma_{\max}$ is the relevant upper volatility bound for the algorithm in Fig. 5.4. Recall that the algorithm computes the discretization in time and space for the explicit finite difference scheme. Mixed explicit/implicit schemes don't require exceptionally small time steps and may in the case of volatility shock scenarios be faster than explicit schemes. For this reason, Crank-Nicholson is used in the following experiments.

The validity of the PDE (4.9) is another numerical issue. Recall that the local volatility under uncertainty is given by

$$\Sigma\left(\frac{\partial^2 f}{\partial S^2}\right) = \begin{cases} \sigma_{\max} & \text{if } \frac{\partial^2 f}{\partial S^2} \geq 0 \\ \sigma_{\min} & \text{if } \frac{\partial^2 f}{\partial S^2} < 0 \end{cases} \tag{9.6}$$

which has the welcome property that $\frac{\partial}{\partial \lambda_n} \Sigma = 0$ almost everywhere, for $1 \leq n \leq |\lambda|$. This has the consequence that the gradient in λ is a solution of (4.9), too, with different boundary conditions. Volatility-shock scenarios have only a finite number of additional transitions in volatility space and therefore do not change this property.

9.1.2 Empirical Results

We compare volatility shock scenarios against classical uncertain volatility scenarios in three experiments with the pricing software Mtg.

Experiment 1: A Butterfly Spread Consider the butterfly spread of four call options in Fig. 9.7. The maturities of the options are 30, 50, 40 and 60 days, respectivly. The current stock price is $S_0 = 100$, and the interest rate is $r = 0.03$.

type	maturity	strike	λ_n
call	30	95	1
call	50	100	-1
call	40	110	-1
call	60	115	1

Fig. 9.7. A butterfly spread consisting of four call options. The spread is not perfect: the maturity dates of the calls are not aligned

The spread is priced under three volatility scenarios:

134   9 Exotic Volatility Scenarios

1. A linear volatility scenario with constant volatility $\sigma = 0.15$.
2. A volatility shock scenario with $\sigma_{min} = \sigma_0 = 0.15$, $\sigma_{max} = 0.5$, duration $d = 3$ days, periodicity $p = 1$ day and frequency $f = 1$.
3. A conventional worst-case volatility scenario with $\sigma_{min} = 0.15$ and $\sigma_{max} = 0.184052$, where the latter was chosen to match the average volatility over any high-volatility path in scenario 2:

$$\sigma_{max} = \sqrt{\frac{1}{60} \int_0^{60} \sigma^2 \, dt} = \sqrt{\frac{1}{60}(3 \times 0.5^2 + 57 \times 0.15^2)} \qquad (9.7)$$

The time step for the Crank-Nicholson finite-difference scheme is $dt = 1/(10 \times 365)$.

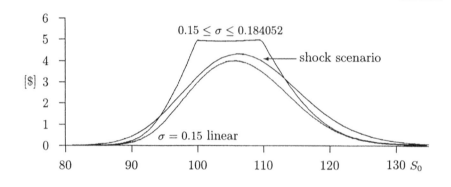

Fig. 9.8. The butterfly spread of Fig. 9.7 priced under three volatility scenarios. Shown is the worst-case value plotted against today's value of the underlying. The parameters for the shock scenarios are $\sigma_{min} = \sigma_0 = 0.15$, $\sigma_{max} = 0.5$, duration $d = 3$, periodicity $p = 1$ and frequency $f = 1$

Figure 9.8 plots the resulting worst-case values against the time-zero value of the underlying. The linear scenario obviously yields the smallest value throughout. The relation between the two non-linear scenarios is less apparent. The volatility shock scenario is smoother and comes closer to the linear scenario. It may be more appealing to practitioners.

Figure 9.9 contains an image of the top-level consolidating lattice instance. Black regions indicate where the maximum operator in step 2(b)i in Fig. 9.5 locates the potential start of a shock period. Conversely, any path starting at time 0 enters its shock period when it hits one of the black regions for the first time. Shock periods are predominantly entered near maturity dates.

Experiment 2: Increasing the Frequency In experiment 2 the portfolio in Fig. 9.7 is priced again, under the same volatility shock scenario with

9.1 Volatility Shocks for Portfolios of Vanilla Options   135

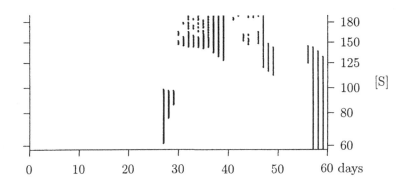

Fig. 9.9. The shock front unveiled. Black regions indicate where three-day shock periods start. The four clusters correspond to the four maturities 30, 40, 50 and 60 days. The spatial axis is in log-scale; the labels are normalized

$\sigma_{\min} = \sigma_0 = 0.15$, $\sigma_{\max} = 0.5$, duration $d = 3$ and periodicity $p = 1$. The frequency f varies between 1 and 20. Figure 9.10 lists the number of lattice instances created and the worst-case value as a function of f. Figure 9.11 shows the worst-case value graphically.

The number of lattice instances created by the dynamic creation scheme is $f \times (\lceil d/p \rceil + 1) + 1 = 4f + 1$. Here, $\lceil d/p \rceil$ is the number of conventional lattice instances per level, $\lceil d/p \rceil + 1$ is the number of overall lattice instances per level, and the additional lattice instance L_0 is used for the level-zero linear pricing. The time complexity is thus linear in the number of lattice instances.

For $f = 20$ the worst-case volatility-shock value and the conventional worst-case value obtained under an uncertain volatility scenario $0.15 \leq \sigma \leq 0.5$ coincide, for 20 volatility shocks of 3 days length each cover the entire 60-day lifetime of the portfolio. A coverage of $10 \times 3 = 30$ days is, according to the data in Fig. 9.10, already sufficient to reproduce the conventional value to within 1.3%.

Figure 9.12 shows the shape of the shock front on the top-level lattice for $f = 2, 3, 4$, respectively. In this context, the top-level lattice for $f = a$ is a lattice instance b levels away from the top for $f = a + b$. If a path hits one of the black regions in the top picture, a three-day volatility-shock period is initiated after which the path continues on the lattice instance shown in Fig. 9.9. Similarly, with intermediate three-day transitions with high volatility oscillation, paths examined under scenarios $f = 3$ and $f = 4$ jump from the middle respectively bottom picture to the top respectively middle picture by passing through one of the black regions. The shock region shows a vertical pattern because shock periods may only start on day boundaries, but the time step is 1/10 of a day.

frequency f	# lattices	value
1	5	3.264
2	9	3.462
3	13	3.648
4	17	3.816
5	21	3.946
6	25	4.051
7	29	4.136
8	33	4.197
9	37	4.240
10	41	4.271
11	45	4.291
12	49	4.306
13	53	4.316
14	57	4.322
15	61	4.325
16	65	4.326
17	69	4.327
18	73	4.327
19	77	4.327
20	81	4.327

Fig. 9.10. Number of lattice instances and worst-case volatility-shock values for the portfolio in Fig. 9.7, as a function of the shock frequency f

Experiment 3: Convergence Experiments 1 and 2 were executed with a time step of $dt = 1/(10 \times 365)$. Worst-case prices for $d = 3$, $p = 1$ and

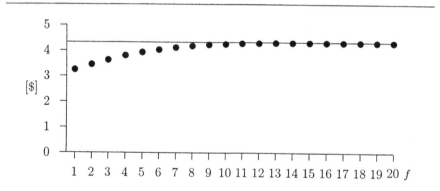

Fig. 9.11. Worst-case volatility-shock values at $S = 100$ for the butterfly spread in Fig. 9.7, as a function of the shock frequency f. The horizontal line represents the worst-case value under the conventional volatility scenario $0.15 \leq \sigma \leq 0.5$

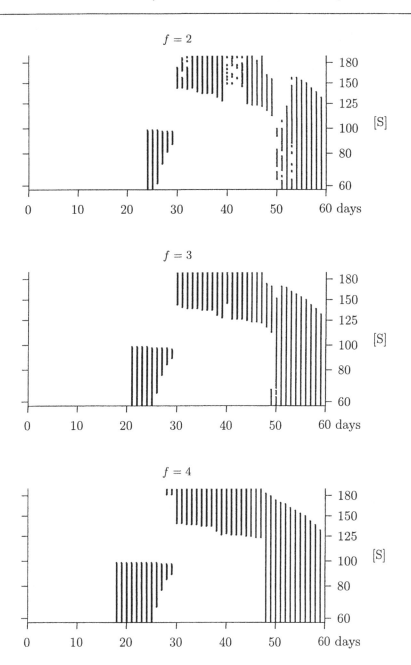

Fig. 9.12. The top-level shock front unveiled for $f = 2, 3, 4$. Black regions indicate where three-day shock periods start. Notice that the shock front expands to the left as the frequency increases. (also compare with Fig. 9.9 for $f = 1$)

$f = 1, 2, 3$, computed with 1, 2, 5, 10, 20, 50 and 100 steps per day and shown in Fig. 9.13, certify the stability of the results obtained. No significant improvement is achieved for $dt < 1/(10 \times 365)$.

time step	price		
	$f = 1$	$f = 2$	$f = 3$
1/365	3.3859	3.5830	3.7672
$1/(2 \times 365)$	3.2922	3.4996	3.6887
$1/(5 \times 365)$	3.2717	3.4732	3.6603
$\rightarrow 1/(10 \times 365)$	3.2637	3.4619	3.6488
$1/(20 \times 365)$	3.2587	3.4571	3.6458
$1/(50 \times 365)$	3.2569	3.4549	3.6437
$1/(100 \times 365)$	3.2560	3.4541	3.6429

Fig. 9.13. Worst-case prices for the call spread in Fig. 9.7 under the shock-volatility scenarios $d = 3$, $p = 1$ and $f = 1, 2, 3$. Results in experiments 1 and 2 were obtained with a time step $dt = 1/(10 \times 365)$. The data shows that there are no convergence issues; doubling the number of steps per day from 10 to 20 changes the result by 0.15, 0.13 and 0.08%, respectively

Conclusion The concept of refined volatility scenario makes direct economic sense. In particular, volatility shock scenarios promise to remedy some of the flaws of conventional uncertain volatility scenarios based on a perpetual volatility band. Among these are too pessimistic price bands and unrealistic mapping of market behavior.

The preceding discussion and experiments prove that the computational overhead is linear in the granularity d/p of the volatility shock scenario, and therefore bearable. No sacrifices have to be made in terms of accuracy.

Figure 9.8 shows that volatility-shock prices are less extremal than prices obtained under conventional uncertain volatility scenarios. Figure 9.11 shows that volatility shock scenarios react gradually to an increase in the extent of volatility oscillation. Volatility shock scenarios therefore permit to fine-tune the market model to a great degree. They promise to be a valuable tool in assessing volatility risk.

9.2 Volatility Shocks and Exotic Options

Exotic volatility scenarios and portfolios of exotic options can be combined. The computational overhead is multiplicative. The algorithm in Fig. 9.4 creates an initial set of lattice instances with signatures $(\mathbf{X}, \lambda, \tau, \xi, \delta)$; additional lattice instances with signatures $(\mathbf{X}', \lambda', \tau', \xi', \delta')$, $(\mathbf{X}', \lambda') \subset (\mathbf{X}, \lambda)$, may be

9.2 Volatility Shocks and Exotic Options

created dynamically later (if the portfolio contains American options) or statically (if the portfolio contains barrier options).

Steps 1 and 2a in Fig. 9.5 refer to the untainted rollback scheme in Fig. 5.5. In the case of exotic options, more sophisticated operations based on dynamic programming need to be executed instead. Chapters 7 and 8 have explained how lattice instances with partial portfolios are maintained to locate the exercise boundary or to supply it with data, if barrier and/or American options are part of the problem. Luckily, this kind of data transfer between lattice instances can be confined to steps 1 and 2a: the maximum operator in the expression for $\hat{V}(j,i;L)$ in step 2(b)i is at the highest level and does not interfer. (In mixed implicit/explicit schemes, however, this may create the same problem as for American options, making iterative refinement of the initial solution of the underlying linear system of equations necessary.)

Figure 9.14 illustrates the distinction between the "horizontal" relationship of lattice instances for different partial portfolios, but with identical volatility shock parameters, and the "vertical" relationship between consolidating and conventional lattice instances with differing volatility shock parameters. The relationship between τ and τ', ξ and ξ', δ and δ' is predetermined and has been discussed above. The relationship between $(\mathbf{X}, \lambda)$ and $(\mathbf{X}', \lambda')$ depends on the makeup of the portfolio. Certain is only that $(\mathbf{X}', \lambda') \subset (\mathbf{X}, \lambda)$.

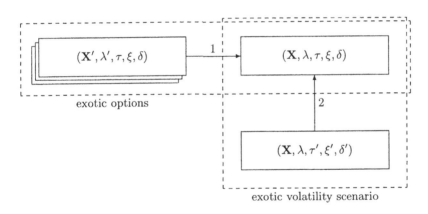

Fig. 9.14. The lattice instance with signature $(\mathbf{X}, \lambda, \tau, \xi, \delta)$ imports data from lattice instances with signatures $(\mathbf{X}', \lambda', \tau, \xi, \delta)$ and $(\mathbf{X}, \lambda, \tau', \xi', \delta')$. The numbers indicate the order in which data is imported

Figure 9.15 generalizes the microscopic example of Fig. 9.14 and shows a data flow diagram for a volatility shock scenario with frequency $f = 2$. Each stack of boxes represents a component scenario for a fixed partial portfolio. The component scenario imports data at every level from a subordinate component scenario located to its left, depending on the requirements arising from the exotic options in the portfolio.

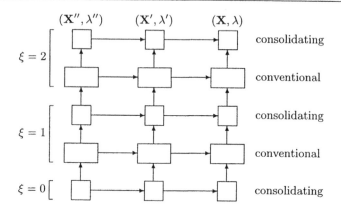

Fig. 9.15. A volatility shock scenario with $f = 2$ for a portfolio **X** containing some exotic options. Each box represents one or more lattice instances; arrows represent the data flow. The size of each box is proportional to the number of lattice instances in the group it represents (in this case, we conjecture $p < d$)

Mtg, our pricer, is capable of handling both exotic options and volatility shock scenarios at the same time.

Part III

Object-Oriented Implementation

10 The Architecture of Mtg

Mtg is an experimental software system for the algorithms of Chapters 7, 8 and 9. It consists of components written in C++ and Java. Figure 10.1 arranges the components of Mtg in a top application layer, and a bottom support layer. Dashed boxes denote third party software.

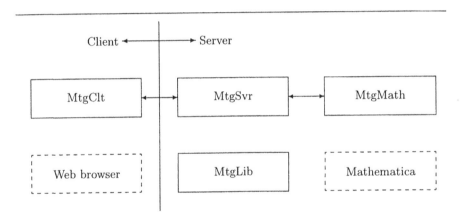

Fig. 10.1. Components MtgSvr and MtgLib are written in C++. MtgClt is written in Java and runs in a Web-browser environment. MtgMath is part C++, part Mathematica script

The components of Mtg are:

MtgLib The core C++ library. MtgLib contains the majority of the code written for this book, about 81500 lines of code. MtgLib is platform-independent.

MtgSvr A background server process. MtgSvr receives and answers requests via TCP. The text protocol used by MtgSvr serves mainly to transmit descriptions of object instances of classes in MtgLib. MtgSvr is a tiny wrapper around MtgLib. Under Unix, MtgSvr is a deamon; on Windows NT, MtgSvr is implemented as a service.

144   10 The Architecture of Mtg

MtgClt A Java front-end that knows how to communicate with MtgSvr. MtgClt can act both as stand-alone application and as applet run by a Web browser. It is powerful enough to let the user solve pricing problems with barrier and American options, under worst-case and volatility shock scenarios. It is, however, restricted to the lattice approach for Black-Scholes. MtgClt consists of approximately 11500 lines of Java code (about half of which is general-purpose).

MtgMath A front-end that uses the symbolic and plotting capabilities of the software system for technical computation, Mathematica. MtgMath was mainly used to do the experiments and prepare the graphs in this book.

Some informal remarks on terminology are in order, predominantly with respect to lattice-based evaluation. *Rollback* is the term used to describe the outer loop that iterates over the time slices $t_N, t_{N-1}, \ldots, t_0$ in the finite difference scheme. The inner loop processing that occurs for each time slice, i.e. the propagation of the solution at time slice t_{i+1} to the earlier time slice t_i, is called rollback *round*. Instead of time slice we sometimes say *hyperplane* to emphasize the data aspect. Under a one-factor model, the hyperplane is actually a two-dimensional plane with rows indexed $s_D, \ldots, s_0, \ldots, s_U$ (see Sect. 5.1), and columns for the total value and each gradient element. The number of columns used is called the *width* of the hyperplane. The *current* round, time slice, hyperplane, or node refers to the current iteration of the rollback loop (forgive the cyclic definition, it should be clear). We use the terms *Monte Carlo* and *simulation* interchangingly, and sometimes together.

11 The Class Hierarchy of MtgLib—External

MtgLib classes that correspond directly to input parameters and have some intuitive "meaning" to the user are called external. Instances of these classes are defined in MtgLib's scripting language MtgScript, as defined in Appendix B. The following categories of external classes exist:

Instruments: Maturity, payoff policy, knock-out policy and early-exercise policy are the dominant orthogonal features of instruments. American/European options with or without knock-out boundaries and with linear of digital payoff are standardized in MtgLib.

Portfolios: Instruments are combined into portfolios which generalize some of their properties (the longest maturity, for instance).

Models: Models consist of specifications of factors and model coefficients, possibly uncertain. A one-factor Black-Scholes model is used in most of this book.

Model coefficients: They may have their own classes to allow term structure. At this time, piecewise constant volatility and drift coefficients are supported. The volatility coefficient may be uncertain.

Scenarios: Models and their (uncertain) coefficients are interpreted according to a prescribed scenario. We have discussed worst-case volatility and volatility scenarios. Some consistency between the model and the scenario is required: if the model incorporates uncertain model coefficients, the scenario must be able to resolve the uncertainty.

Numerical methods: MtgLib supports explicit or mixed implicit/explicit finite difference schemes as well as Monte-Carlo simulation. The requirements diverge: while finite difference schemes are based on a collection of lattice instances, Monte Carlo methods require path instances which are treated differently. In MtgLib, lattices and path spaces are indeed separate objects.

Evaluators: At first objects are passively stored in a repository. Concrete pricing operations are only triggered by specifying an evaluator object which lives only while the particular portfolio/model/scenario combination is evaluated. Evaluators format and output the result.

Figure 11.1 shows an example script that, when submitted to the program MtgSvr, initiates the computation of the worst-case price of a portfolio of three puts and one call, under a volatility scenarion $0.1 \leq \sigma \leq 0.2$. The script contains instances of all the classes listed above.

```
1  claim a {
2      type american_put, maturity 30, strike 100 }
3  claim b {
4      type european_put, maturity 25, strike 100 }
5  claimc {
6      type european_call, maturity 20, strike 100,
7      up-and-out 110 }
8  claim d {
9      type european_put, maturity 15, strike 100,
10     down-and-out 90 }
11
12 portfolio p { a long 200, b short 10, c long 2, d short 1 }
13
14 factor s {}
15 vol v { implied 30 10%..20% }
16 drift r { implied 30 2.5% }
17 model m { type back_scholes, vol v, discount r, s 100 }
18 scenario s { type worst_case, seller }
19
20 lattice l { model m, portfolio p, tree 3.5, time_step 0.5 }
21
22 evaluate { model m, lattice l, scenario s, portfolio p }
```

Fig. 11.1. An example of MtgScript, consisting of a portfolio of three puts and one call

The following sections discuss each class category in more detail. Although code fragments are included, this overview is not a tutorial on how to use MtgLib. Instead, it emphasizes design principles.

For historical reasons, it also emphasizes lattice-based evaluation based on worst-case and volatility shock scenarios over Monte-Carlo and minimum-entropy calibration. Initially, MtgLib exclusively supported lattice-based evaluation; Monte-Carlo methods followed later.

11.1 Instruments

The class hierarchy into which instruments are organized is shown in Fig. 11.2. The parent class tClaim is abstract and needs to be instantiated in subclasses.

Class name	Purpose
tClaim	Parent class (abstract)
tStdClaim	Standard calls and puts
tCustomClaim	Customizable in a mini-language
tUSTBond	US-Treasury bond
tCashflow	Supporting class (abstract)

Fig. 11.2. The hierarchy of instrument classes. Indentation indicates inheritance. Standard instruments are calls and puts, American or European, with linear or digital payoff, with or without barriers. US Treasury bonds are also supported

tClaim provides a unified interface to relevant instrument properties. Its definition is shown in Fig. 11.3. The scalar properties listed in the private section are initialized from script declarations common for all instrument types. The virtual functions in the public section must be overridden in subclasses to create the unique outlook of the particular instrument type. The middle section contains two functions that are used during the construction of the finite difference lattice: getEvents() must deliver the location of all relevant events (maturity, cashflow, barrier, early exercise or otherwise) on the time axis. The lattice is then guaranteed to match these events. getBarriers() may (but is not forced to) return the location in space of eventual knock-out barriers. Designing the lattice to match those increases numerical accuracy, but is not mandatory.

The semantic of the member variables and functions is summarized in the following paragraphs.

m_nMaturity indicates the number of days to maturity. MtgLib does not support real calendar dates for option types (the subclass tUSTBond does contain a calendar maturity date in addition, though).
m_gMultiplier represents the position in the instrument, mathematically λ.
m_bMonitor is a boolean flag that indicates whether the virtual member function monitor() should be used or not. This flag is true for American options.
m_Cashflow is a list of objects derived from the abstract class tCashflow, whose definition is given in Fig. 11.4. Each cashflow object implements an additional, possibly space-dependent cashflow on a fixed date.
payoff() computes the payoff at maturity. The tEngine object whose reference is passed to payoff() and all other functions in tClaim provides information about the current state. For lattice instances, the Engine contains the current node instance $(j, i; L)$ on which payoff() must base its calculation. Assuming a one-factor model, for example, the values of s_j and t_i can be queried with Engine.day() and Engine.factor(), respectively. Engines are discussed below, in Sect. 12.1.

```
 1 class tClaim : public tObject {
 2
 3     int m_nMaturity;
 4
 5     double m_gMultiplier;
 6
 7     bool m_bHasUpBarrier;
 8     double m_gUpBarrier;
 9
10     bool m_bHasDownBarrier;
11     double m_gDownBarrier;
12
13     bool m_bMonitor;
14
15     tCashflow* m_Cashflow[...];
16
17 protected:
18
19     virtual void getEvents( ... ) const;
20     virtual void getBarriers( ... ) const;
21
22 public:
23
24     virtual double payoff( tEngine& Engine );
25     virtual double knockoutPayoff( tEngine& Engine );
26     virtual double exercisePayoff( tEngine& Engine );
27
28     virtual bool upBarrier( tEngine& Engine,
29         double& gBarrier );
30     virtual bool downBarrier( tEngine& Engine,
31         double& gBarrier );
32
33     virtual tExPolicy monitor( tEngine& Engine,
34         double gUnitValue );
35 };
```

Fig. 11.3. A crude sketch of the class definition of tClaim. Possible values of the enumeration type tExPolicy are DontExercise, ForceExercise and MayExercise

knockoutPayoff() computes the premium at knock-out. Unless overridden, this function always returns 0.

exercisePayoff() computes the payoff received at early exercise. By default, this function calls and returns the result of payoff().

11.1 Instruments   149

upBarrier() returns true if there is an up-and-out barrier for the current time slice, as determined by Engine. If also returns the barrier itself. Unless overridden, upBarrier() is defined as

```
1  bool tClaim::upBarrier( tEngine& Engine,
2          double& gBarrier ) {
3      if( m_bHasBarrier ) {
4          gBarrier = m_gUpBarrier;
5          return true;
6      }
7      return false;
8  }
```

downBarrier() works in an analoguous way for down-and-out barriers.
monitor() returns a safe estimate (!) of the local early-exercise policy. Possible return values are
- DontExercise if the instrument must not be exercised under the current state.
- ForceExercise if the instruments must be exercised at once.
- MayExercise if the instrument may or may not be exercised. Any further decision depends on the outlook for the entire portfolio and cannot be determined by the instrument in isolation.

Note the analogy with the concepts of continuation, exercise and corridor of uncertainty developed in Chapter 8. It is *not*, however, the task of monitor() to implement any of the speed-up techniques of Sect. 8.2. This is done by the compute engine in cooperation with the scenario object (see Sect. 11.5). The proper implementation for standard American options is thus simply

```
1  tExPolicy monitor( tEngine& Engine,
2          double gUnitValue ) {
3      return MayExercise;
4  }
```

monitor() can also be used to implement irregular barriers, bypassing the upBarrier() and downBarrier() member functions. In this case, continuation and knock-out regions are deterministic. They are implicitly located through the distribution of DontExercise and ForceExercise return values.

The definition of the supporting class tCashflow is given in Fig. 11.4. The member array m_Cashflow is examined during rollback just like the functions upBarrier() and downBarrier() are called for each time slice. The simplest instantiation of tCashflow would override the generate() member function with

```
1  double generate( tEngine& Engine ) {
2      return c;
3  }
```

where c is some fixed coupon payment.

```
1 class tCashflow {
2
3     int m_nDay;
4
5 protected:
6
7     virtual double generate( tEngine& Engine );
8 };
```

Fig. 11.4. The definition of abstract class `tCashflow`. Cashflows are generated on day boundaries

`tStdClaim` instantiates `tClaim` and supports instruments with the following orthogonal features:

- Call or put option?
- Linear or digital payoff?
- American or European?
- Up-and-out and/or down-and-out barrier?

Strike and maturity are the remaining properties.

`tCustomClaim` also instantiates `tClaim`, but does so in a customizable manner by parsing flexible script expressions for

- The payoff at maturity,
- the payoff at knock-out (if relevant),
- the payoff at early exercise (if relevant),
- the location of the knock-out barrier (time-dependent),
- a policy for determining early exercise,
- optional cashflows at fixed dates based on `tCashflow`.

Figure 11.5 shows how these expressions are embedded into the parent class `tClaim`. The definitions of the classes `tNumericalExpr` and `tExPolicyExpression` are not shown here; we merely note that both classes provide a member function `apply()` which is used to evaluate the expression. `payoff()`, for instance, uses a `tNumericalExpr` as follows:

```
1 double tCustomClaim::payoff( tEngine& Engine ) {
2     if( m_pPayoff != 0 )
3         return m_pPayoff->apply( Engine );
4     return 0;
5 }
```

Both tNumericalExpr and tExPolicyExpression objects are constructed from MtgScript expressions. Expressions have access to the state information contained in Engine through keywords such as time. The following script fragment, for instance, defines an up-and-out barrier call with strike 110 and barrier 120, where the barrier is only active for the first 50 days after settlement.

```
1 claim x {
2     type custom, maturity 100,
3     payoff { max( s - 110, 0 },
4     up_and_out { if time < 50 then 125 endif }
5 }
```

tUSTBond also instantiates tClaim. It is modelled after US-Treasury bills and bonds and can compute accrued interest as well as convert between clean and dirty prices.

11.2 Portfolios

Portfolios are collections of instruments. As such, they provide a generalized interface to some of the properties of instruments. The class tPortfolio is final; there are no subclasses.

A definition is given in Fig. 11.6. The meaning of the individual class members is as follows:

m_Claim References to all claims are collected here.

m_Factor In a multi-factor setting, different instruments may refer to different factors, or to the same factors in a different order. To establish a unique order of factors, the factors referenced in any of the instruments are collected in the array m_Factor.

maturity() The longest maturity of any of the instruments in the portfolio.

claim() To access individual instruments, this function must be used. The argument refers to the position of the instrument in m_Claim, which is sorted by maturity.

getEvents() This function in turn calls tClaim::getEvents() for each instrument and amalgamates the result, which is used in another place to calculate the discretization of the time axis for the lattice.

getBarriers() Calls tClaim::getBarriers() for each instrument and combines the result, which is used by the algorithm in Fig. 5.4 to place the spatial levels of the lattice.

matchFactors() The factors in m_Factor are collected without knowledge of the particular model under which the portfolio is to be evaluated. If the elements of m_Factor are the factors $S_1, \ldots, S_n$, and the model makes use of factors $S'_1, \ldots, S'_m$, then $n = m$ must be asserted and the correct mapping found. This task is done by matchFactors().

```
 1 class tCustomClaim : public tClaim {
 2
 3     class tCustomCashflow : public tCashflow {
 4     public:
 5         tNumericalExpr* m_pExpr;
 6         double generate( tEngine& Engine );
 7     };
 8
 9     tNumericalExpr *m_pPayoff;
10     tNumericalExpr *m_pKnockoutPayoff;
11     tNumericalExpr *m_pExercisePayoff;
12
13     tNumericalExpr *m_pUpBarrier;
14     tNumericalExpr *m_pDownBarrier;
15
16     tExPolicyExpr *m_pMonitor;
17
18     double payoff( tEngine& Engine );
19     double knockoutPayoff( tEngine& Engine );
20     double exercisePayoff( tEngine& Engine );
21
22     bool upBarrier( tEngine& Engine,
23         double& gBarrier );
24     bool downBarrier( tEngine& Engine,
25         double& gBarrier );
26
27     tExPolicy monitor( tEngine& Engine,
28         double gUnitValue );
29 };
```

Fig. 11.5. The definition of tCustomClaim, an instantiation of tClaim. The corresponding extension of tCashflow is defined locally

The functionality of portfolio objects is mostly used in the preparatory stage of evaluation. During actual rollback or simulation, instruments are directly accessed through the claim() member function.

11.3 Models

Just like instruments, models are supported through an abstract parent class, tModel, and child classes which provide the model-specific body. Figure 11.7 shows the dependencies.

```
1 class tPortfolio {
2
3     tClaim* m_Claim[...];
4     tFactor* m_Factor[...];
5
6 public:
7
8     int maturity() const;
9     tClaim& claim( int nPos ) const;
10
11    void getEvents( ... ) const;
12
13    void getBarriers( const tFactor* pFactor,
14        double Barrier[...] ) const;
15
16    tRetCode matchFactors( const tModel& Model ) const;
17 };
```

Fig. 11.6. A very condensed definition of tPortfolio

By definition, models are independent of the numerical method under which they are evaluated. In practice, models must provide services to the actual numerical method used. In MtgLib, models are constructed by the user without reference to a numerical method, but may chose to cooperate or defect later when the evaluator object tries to match the model and the numerical method.

The child class tBSModel is only used for lattice-based evaluation under the Black-Scholes model. The child classes tHJMGaussianModel and tVasicekModel support only Monte Carlo methods for fixed income instruments.

Class name	Purpose
tModel	Parent class (abstract)
tBSModel	One-factor Black-Scholes model
tHJMGaussianModel	Multi-factor HJM
tShortRateModel	Parent class (abstract)
tVasicekModel	A mean-reversion model

Fig. 11.7. The model hierarchy. Fixed income models are parenthesized because our main interest is in equity/FX models on a lattice

```
 1  class tModel {
 2
 3      tFactor* m_Factor[...];
 4      tCalendar* m_pCalendar;
 5
 6  public:
 7
 8          // Functions for lattice-base methods:
 9
10      virtual tRetCode createEngine(
11          const tScenario* pScenario,
12          tFDEngine*& pEngine, tAccuracy m_nAccuracy );
13
14      virtual tRetCode createSpaceAxis( tFDMethod nMethod,
15          double gMaxDt, tSpaceAxis* Space[...],
16          const tPortfolio* pPf = 0 );
17
18          // Functions for simulation methods:
19
20      virtual tRetCode createEngine( tMCEngine*& pEngine );
21
22      virtual tRetCode createEvolution(
23          const tPathSpace& PathSpace,
24          tMCEngine::tEvolutionStub*& pEvolution ) const;
25  };
```

Fig. 11.8. The crucial members of class tModel. A model that supports lattice-based methods must implement the first two virtual functions (FD = finite differences). A model that supports simulation methods must implement the last two virtual functions (MC = Monte Carlo)

The definition of the class tModel is shown in Fig. 11.8. The semantics of the member components of tModel are as follows:

m_Factor The number of factors in the model is not predetermined. Factors are added when the model object is created. However, the number of factors must be known at the level of the parent class tModel (functions to query the number of factors and other trivial information are not included in the figure). For this reason, references to factors are stored in m_Factor.

m_pCalendar The calendar object is optional and at this time only supplies the scaling factor for the conversion between day and year-based quantities: the time-unit used in lattice calculations is one day, while model coefficients are usually quoted in their annualized form. If no calendar object is specified, a year of 365 days is assumed.

Before we proceed to describe public member functions, a remark on compute engines. The knowledge about the factor dynamics is encapsulated in the model. It provides information about PDE's to lattice-based methods and SDE's to simulation methods. It does so by creating model-specific compute engines, which are all based on either tFDEngine or tMCEngine (in turn based on the parent class tEngine).

Engines are discussed in more detail below, in Sect. 12.1. At this point we emphasize that different types of engines are created for lattice-based and simulation-based computation: tFDEngine and tMCEngine are the respective child classes.

createEngine(), first version Creates a compute engine for lattice-based evaluation. To create the proper engine, the model must know the scenario. Volatility shock scenarios, for instance, make lattice instance creation more complicated than worst-case volatility scenarios (see Chapter 9). It is the engine which creates and maintains lattice instances.

The model must also know the selected speed-up technique for American options. This parameter, nAccuracy, is forwarded to the created engine. The name nAccuracy reflects the generality of the parameter; the engine chooses the speed-up technique that matches the parameter. Possible values are Exact (corresponding to the maintainance of corridors of uncertainty) and Low (corresponding to the collapsing of corridors of uncertainty).

createSpaceAxis() Creates the spatial discretization for lattice-based models, based on the algorithms in Figs. 5.4 and 5.5, and returns it in Space. Space is an array with one entry per factor. The parameter nMethod can take the values Explicit and Implicit and controls to what extent stability is a concern. gMaxDt corresponds to the input parameter dt_{max} in Fig. 5.4. The optional parameter pPf references a portfolio object. If present, the portfolio barriers are retrieved with pPf->getBarriers() and passed to the algorithm in Fig. 5.4.

createEngine(), second version This version creates a compute engine for simulation methods. At this point, simulation methods are not scenario based and therefore no additional arguments are required.

createEvolution() Simulation methods work by shooting one random path at a time in the so called "path space." The random path is then converted into the corresponding factor paths by calling createEvolution(). Evolutions are organized in their own separate class hierarchies, not shown here.

One possible instantiation of tModel is shown in Fig. 11.9. tBSModel only supports lattice-base evaluation for one-factor Black-Scholes with time-

```
 1 class tBSModel : public tModel {
 2
 3     tDrift* m_pDiscount;
 4     tDrift* m_pCarry;
 5     tDrift* m_pMu;
 6
 7     tVol* m_pVol;
 8
 9     double m_gRoot;
10
11 public:
12
13     tRetCode createSpaceAxis( tFDMethod nMethod,
14         double gMaxDt,
15         tSpaceAxis Space[...], const tPortfolio* pPf = 0 );
16
17     tRetCode createEngine( const tScenario* pScenario,
18         tFDEngine*& pEngine, tAccuracy nAccuracy );
19 };
```

Fig. 11.9. The class tBSModel, the model body for one-factor Black-Scholes. Only lattice-based evaluation is supported

varying coefficients. The member components have the following interpretation:

m_pDiscount References an interest-rate term structure for the parameter r of (4.9). The class tDrift is explained in Sect. 11.4.

m_pCarry References a term structure for the dividend rate or foreign interest rate, depending on whether the underlying asset is a stock or an exchange rate.

m_pMu The no-arbitrage drift parameter. References a term structure for the difference between m_pDiscount and m_pCarry.

m_pVol References a volatility term structure that may exhibit uncertainty. The object *m_pVol contains upper and lower bounds for the local volatility for each time slice.

m_gRoot The initial value S_0 of the underlying asset. The lattice is constructed such that $s_0 = S_0$.

createSpaceAxis() Finds the stable spatial discretization that matches the barriers of the (optional) portfolio parameter and returns exactly one object instance of the class tGeoSpaceAxis. The prefix "Geo" means geometric Brownian motion. The member function prepare() of this class, called in createSpaceAxis(), uses the algorithm in Fig. 5.4. Figure 11.10 contains a skeleton of createSpaceAxis().

createEngine() Creates the lattice-based compute engine appropriate for the scenario parameter. Two types of compute engines for one-factor Black-Scholes are currently implemented: tGeoEngine for worst-case volatility scenarios (class tWorstCase) and tShockEngine for volatility shock scenarios (class tShockScenario). The function is outlined in Fig. 11.11.

(Remark: the way createEngine() is coded leads to an extensibility problem. Prolongued sequences of conditional statements guarded by dynamic down-casts should be avoided. There are a handful of spots in MtgLib where this problem occurs.)

```
1  tRetCode tBSModel::createSpaceAxis( tFDMethod nMethod,
2      double gMaxDt, tSpaceAxis Space[...],
3      const tPortfolio* pPf )
4
5  {
6      double gMinVol, gMaxVol, gMinMu, gMaxMu;
7
8      m_pVol->getFwdRange( gMinVol, gMaxVol );
9      m_pMu->getFwdRange( gMinMu, gMaxMu );
10
11     tGeoSpaceAxis* p = new tGeoSpaceAxis;
12
13     if( pPf != 0 ) {
14         double Barrier[...];
15
16         pPf->getBarriers( m_Factor[0], Barrier );
17         p->prepare( nMethod, m_gRoot, gMinVol, gMaxVol,
18                     gMinMu, gMaxMu, Barrier );
19     }
20     else {
21         p->prepare( nMethod, m_gRoot, gMinVol, gMaxVol,
22                     gMinMu, gMaxMu );
23     }
24
25     Space.append( p );
26     return OK;
27 }
```

Fig. 11.10. A sketch of the member function tBSModel::createSpace(). The class tGeoSpaceAxis is derived from tSpaceAxis and supports geometric Brownian motion models. It calls tGeoSpaceAxis::prepare(), which implements the algorithm in Fig. 5.4

```
 1  tRetCode tBSModel::createEngine( const tScenario* pScenario,
 2      tFDEngine*& pEngine, tAccuracy nAccuracy )
 3
 4  {
 5      if( dynamic_cast<const tShockScenario*>(
 6              pScenario ) != 0 ) {
 7          pEngine = new tShockEngine;
 8      }
 9      else
10      if( dynamic_cast<const tWorstCase*>(
11              pScenario ) != 0 ) {
12          pEngine = new tGeoEngine;
13      }
14      else {
15          return NOT_AVAILABLE;
16      }
17
18      pEngine->setAccuracy( nAccuracy );
19      return OK;
20  }
```

Fig. 11.11. The member function tBSModel::createEngine(). tShockEngine and tGeoEngine are both derived from tFDEngine. Both classes support one-factor geometric Brownian motion models (the "Geo" prefix)

11.4 Model Coefficients

Constant model coefficients such as the initial value of the underlying asset are stored by the model class directly: tBSModel::m_gRoot is an example. Other model coefficients have more structure and deserve their own classes. As coefficients depend on the actual model, corresponding classes may be quite diverse, and form a collection rather than a strictly hierarchical class tree.

Figure 11.12 shows the inheritance relations for the model coefficient classes currently supported in MtgLib. The tTermStruct hierarchy was developed first and does not support real calendar dates; real calendar dates are only handled on the level of tDrift and tVol and below. The classes tHJMTermStruct and tShortRateTermStruct, on the other hand, were developed with support for real calendar dates already in mind. They are used as model coefficient classes for interest-rate models.

11.4 Model Coefficients

Class name	Purpose
tTermStruct	Parent term structure class (abstract)
tLinTermStruct	Term structure for linear coefficient
tSqTermStruct	Term structure for quadratic coeffiient
tDrift	General term structure class for drift (abstract)
tStepDrift	Piece-wise constant drift
tVol	General term structure class for volatility (abstract)
tStepVol	Piece-wise constant volatility
tHJMTermStruct	For multi-factor HJM
tShortRateTermStruct	For models like Vasicek

Fig. 11.12. Classes for model coefficients. These classes are model-dependent; they do not share a common base class

11.4.1 The Base Class tTermStruct

tTermStruct is the core class for piece-wise constant time-varying model coefficients. Let α be a model coefficient for an n-factor model with factors $X_1, \ldots, X_n$. If

$$\alpha(X_1(t), \ldots, X_n(t), t) = \alpha(t) = c_k \tag{11.1}$$

with k such that $t \in (t_{k-1}, t_k]$ for some times slices t_{k-1} and t_k, then α can be modelled by tTermStruct.

The basic functionality of tTermStruct is to compute the values

$$\alpha(t) \tag{11.2}$$

and

$$\frac{1}{b-a} \int_a^b \alpha(t)\, dt \tag{11.3}$$

fast. The granularity of t, a and b is assumed to be one day.

tTermStruct can be used for both linear and quadratic parameters, such as drift and volatility. This generality is achieved by introducing a scaling function ϕ and replacing (11.3) with

$$\phi^{-1}\left(\frac{1}{b-a} \int_a^b \phi(\alpha(t))\, dt\right) \tag{11.4}$$

For linear term structures, ϕ is the identity and $\alpha(t)$ the forward rate. For quadratic term structures such as volatility, $\phi(x) = x^2$.

Assume jumps occur at jump points $u_0, u_1, \ldots, u_M$, where each u matches some time slice t_i. To compute (11.2) and (11.4) fast, the following quantities are maintained for each jump point u:

$$\begin{aligned} \texttt{m\_gFwd} &= \alpha_u \\ \texttt{m\_gFwd2} &= \phi(\alpha_u) \\ \texttt{m\_gImp} &= \int_0^u \phi(\alpha(t))\, dt \end{aligned} \qquad (11.5)$$

(11.4) can then be computed for all intermediate time slices t with $O(1)$ overhead, by using m_gImp of the previous sample point u as a base and adding m_gFwd2, multiplied by the number of days between u and t. Subsequent normalization is straightforward.

The remaining problem is to locate the previous jump point u for a given intermediate time slice t, if t is not also a jump point. A shallow forest of bounded depth does the trick for tTermStruct. At the highest level of the forest, each node represents 100 consecutive days. Only if such a period contains a jump point is refinement necessary: the corresponding node branches into 10 child nodes, each covering a period of 10 days, and so on. The memory requirements for this data structure are still linear in the number of days covered by the term structure, but nevertheless reduced 100-fold compared to day-by-day storage if the number of jumps is small.

Figure 11.13 summarizes the important components of class tTermStruct. Their interpretation is as follows:

m_Spec The nested type tSpec describes one jump point. The member variable m_nUnit locates the jump point in time. m_gFwd, m_gFwd2 and m_gImp are defined in (11.5). tMap is a template for the efficient implementation of the shallow forest data structure mentioned above. m_Spec is built as jump points are added with addForward() and addImplied().
scaleUp() The scaling function ϕ.
scaleDown() The inverse of the scaling function ϕ.
addForward() and addImplied() The term structure object is constructed by calling addForward() or addImplied() for each jump point. When all jump points have been specified, the constant term structure coefficient between each pair of jump points is determined as follows:
- If the jump point $u = t_k$ has been added with addForward(), then the coefficient between the previous jump point and u is simply set to the actual value of the gFwd parameter.
- If the jump point $u = t_k$ has been added with addImplied(), then the coefficient between the previous jump point and u is chosen such that the integral matches gImp, i.e.

$$\phi^{-1}\left(\frac{1}{u}\int_0^u \phi(\alpha(t))\, dt\right) = \texttt{gImp}$$

Note that this calculation can fail if ϕ is not the identity!
The parameters nUnit, nFromUnit and nToUnit indicate the endpoint of the respective time unit, where time units are counted from zero.

11.4 Model Coefficients

```
 1 class tTermStruct {
 2
 3     struct tSpec {
 4         int m_nUnit;
 5         double m_gFwd;
 6         double m_gFwd2;
 7         double m_gImp2;
 8     };
 9
10     tMap<tSpec> m_Spec;
11
12 protected:
13
14     virtual double scaleUp( double gFwd ) const;
15     virtual double scaleDown( double gFwd ) const;
16
17 public:
18
19     tRetCode addForward( int nMaturity, double gFwd );
20     tRetCode addImplied( int nMaturity, double gImp );
21
22     void getFwdRange( double& gMin, double& gMax ) const;
23
24     double forward( int nUnit ) const;
25     double forward( int nFromUnit, int nToUnit ) const;
26
27     int constantUntil() const;
28     int certainUntil( const tTermStruct& TS ) const;
29 };
```

Fig. 11.13. The skeleton of class tTermStruct. Although the basic time unit is one day in most cases, the code itself is independent of the concrete time unit. m_nUnit, nUnit, nFromUnit and nToUnit are therefore used instead of m_nDay, nDay, nFromDay and nToDay

getFwdRange() Once all jump points have been added and the term structure has been finalized, the oscillation of the term structure can be determined by calling getFwdRange(). This information is important for the construction of a stable lattice under the explicit finite difference scheme.

forward() The one- and two-parameter versions correspond to (11.2) and (11.4), respectively.

162   11 The Class Hierarchy of MtgLib—External

constantUntil() It may be useful to know the length of the initial constant segment of the term structure. constantUntil() returns this information.

certainUntil() It is also useful to know wether a certain configuration of model coefficients exhibits uncertainty at all. certainUntil() tests this by comparing the current term structure with the argument TS, and returning the length of the initial segment on which they agree.

The term structure extrapolates beyond the first and last jump points by propagating the values of the first and last constant term structure segment to $-\infty$ and ∞.

11.4.2 Classes Derived from tTermStruct

The child class tLinTermStruct is a straightforward instantiation of tTermStruct, with ϕ being the identity. The child class tSqTermStruct is mildly more complicated; Figure 11.14 shows its definition, together with the implementation of the scaling function and its inverse.

```
 1  class tSqTermStruct : public tTermStruct {
 2
 3      double scaleUp( double gFwd ) const {
 4          return gFwd * gFwd;
 5      }
 6
 7      double scaleDown( double gFwd ) const {
 8          return sqrt( gFwd );
 9      }
10  };
```

Fig. 11.14. The child class tSqTermStruct. Shown are both declaration and definition of the scaling function and its inverse

11.4.3 Classes with tTermStruct Components

The classes tDrift and tVol are more interesting. They are independent of any actual implementation of the drift or volatility term structure and offer a standard query interface for forward rates or volatilities. Piece-wise constant realizations of drift or volatility term structure are obtained by combining the tDrift or tVol shell with tTermStruct as "meat." The formula is

$$\text{tDrift} + \text{tLinTermStruct} = \text{tStepDrift}$$

11.4 Model Coefficients

and

$$\text{tVol} + 2 \times \text{tSqTermStruct} = \text{tStepVol}$$

tLinTermStruct and tSqTermStruct contribute as member components. The class tStepDrift contains a tLinTermStruct object and forwars queries to it; tStepVol contains two tSqTermStruct objects to allow for uncertainty, and forwards queries to them.

```
1  class tDrift {
2
3  public:
4
5      tDrift();
6
7      virtual void getFwdRange( double& gMin,
8          double& gMax ) const;
9
10     virtual double forward( int nUnit ) const;
11     virtual double forward( int nFromUnit,
12         int nToUnit ) const;
13
14     virtual double implied( int nMaturity ) const;
15
16     virtual int constantUntil() const;
17 };
```

Fig. 11.15. Class tDrift is an abstract interface for drift coefficients. Piece-wise constant drift term-structures are one possible instantiation of tDrift

Figure 11.15 shows the abstract tDrift interface. All virtual functions are pure. A call to implied() is equivalent to a call to forward() with the first parameter set to zero.

Figure 11.16 shows the tVol interface. Again, all virtual functions are pure. However, any concrete instantiation of tVol is expected to initialize the following member variables correctly:

m_nConstantUntil The length of the initial period during which the volatility is constant (there is no uncertainty during that period!).

m_nCertainUntil The length of the initial period during which the volatility is certain. Necessarily, m_nConstantUntil $\leq$ m_nCertainUntil.

Both m_nConstantUntil and m_nCertainUntil can be retrieved from the object with trivial functions not shown in the figure.

```
 1 class tVol {
 2
 3     int m_nConstantUntil;
 4     int m_nCertainUntil;
 5
 6 public:
 7
 8     virtual void getFwdRange( double& gMin,
 9         double& gMax ) const;
10
11     // Return a single value:
12     virtual double forward( int nUnit ) const;
13     virtual double forward( int nFromUnit,
14         int nToUnit ) const;
15
16     virtual double implied( int nMaturity ) const;
17
18     // Return a range of value:
19     virtual void forward( int nUnit, double& gMin,
20         double& gMax ) const;
21     virtual void forward( int nFromUnit, int nToUnit,
22         double& gMin, double& gMax ) const;
23
24     virtual void implied( int nMaturity, double& gMin,
25         double& gMax ) const;
26 };
```

Fig. 11.16. Class tVol is an abstract interface for volatility coefficients. Piece-wise constant volatility term-structures are one possible instantiation of tVol

tVol has two sets of volatility retrieval functions: the first set returns a single value, the second a range of values in reference parameters gMin and gMax. If there is no uncertainty, both versions are equivalent. If there is uncertainty, however, and the volatility bounds differ, then only the second set of retrieval functions is required to return the original volatility bounds defined during construction of the object faithfully. The first set of retrieval functions may return any value, the arithmetic average between the minimum and maximum being one example and some additional prior volatility another.

It was mentioned earlier that tDrift and tVol are capable of handling real calendar dates. These features are omitted in Figs. 11.15 and 11.16.

```cpp
1  class tStepDrift : public tDrift {
2
3      tLinTermStruct m_TermStruct;
4
5  public:
6
7      void getFwdRange( double& gMin, double& gMax ) const {
8          m_TermStruct.getFwdRange( gMin, gMax );
9      }
10
11     double forward( int nUnit ) const {
12         return m_TermStruct.forward( nUnit );
13     }
14
15     double forward( int nFromUnit, int nToUnit ) const {
16         return m_TermStruct.forward( nFromUnit, nToUnit );
17     }
18
19     double implied( int nMaturity ) const {
20         return m_TermStruct.implied( nMaturity );
21     }
22
23     int constantUntil() const {
24         return m_TermStruct.constantUntil();
25     }
26 };
```

Fig. 11.17. Piece-wise linear drift coefficients are of type tStepDrift, essentially based on the functionality of tLinTermStruct

Piece-wise linear drift and volatility coefficients are finally realized in the classes tStepDrift and tStepVol ("Step" for step function). Figure 11.17 shows the definition and implementation of tStepDrift, which basically acts as a proxy for its tLinTermStruct member object. Figure 11.18 shows the definition of tStepVol, whose implementation is only slightly less trivial. The two-parameter forward() function, for instance, is implemented as

```cpp
1  void tStepVol::forward( int nFromUnit, int nToUnit,
2      double& gMin, double& gMax ) const
3
4  {
5      gMin = m_MinTermStruct.forward( nFromUnit, nToUnit );
6      gMax = m_MaxTermStruct.forward( nFromUnit, nToUnit );
7  }
```

```cpp
class tStepVol : public tVol {

    tSqTermStruct m_MinTermStruct;
    tSqTermStruct m_MaxTermStruct;

public:

    void getFwdRange( double& gMin, double& gMax ) const;

    double forward( int nUnit ) const;
    double forward( int nFromUnit, int nToUnit ) const;
    double implied( int nMaturity ) const;

    void forward( int nUnit, double& gMin,
        double& gMax ) const;
    void forward( int nFromUnit, int nToUnit,
        double& gMin, double& gMax ) const;

    void implied( int nMaturity, double& gMin,
        double& gMax ) const;
};
```

Fig. 11.18. Piece-wise linear, possibly uncertain volatility coefficients are of type tStepVol. The implementation of the member functions is almost as trivial is those of tStepDrift

The objects v and r in the script shown in Fig. 11.1 are automatically implemented as tStepVol and tStepDrift instances, respectively. tStepVol and tStepDrift instances are used to model the volatility, interest rate, and dividend rate or foreign exchange rate in class tBSModel (see Fig. 11.9).

11.5 Scenarios

Scenario objects perform an "advisory" function for lattice-based evaluation. They are used by compute engines derived from tFDEngine to determine locally how to select the uncertain model coefficients. They also control the assignment of lattice nodes to continuation and exercise regions, and the corridor of uncertainty. Figure 11.19 shows the class hierarchy.

Class name	Purpose
tScenario	Parent class (abstract)
tWorstCase	Worst-case volatility scenario
tShockScenario	Volatility shock scenario

Fig. 11.19. Scenario classes. Each class extends the functionality of its parent

11.5.1 The Base Class tScenario

Figure 11.20 shows how these two tasks are reflected in the class definition. Before indiviual components are discussed, however, we need to clarify the usage of *tags*.

In Defs. 4 and 23, the notion of (extended) lattice signatures has been introduced to uniquely identify individual lattice instances in the collection of lattice instances maintained for the particular pricing problem. To be able to handle a wide variety of scenarios types, a concrete scenario implementation must provide a way to translate a scenario-dependent lattice signature $(\mathbf{X}, \lambda, \text{optseq})$ into a regularized signature which is easier (and faster) to process.

Definition 25 (Regularized lattice signature). *Let $(\mathbf{X}, \lambda, \text{optseq})$ be the pattern of signatures for lattice instances for a concrete scenario, with optseq denoting an optional sequence of coefficients. Then triples of the form $(\mathbf{X}, \lambda, 2m)$, $m \geq 0$, are called* regularized lattice signatures *for the scenario if there is an unambiguous mapping between the two patterns. Furthermore, for the root lattice instance from which the final result is retrieved, $2m = 0$ must hold. $2m$ is called the* signature *tag.*

Compute engines use this regularized form to manage the storage of lattice instances. Regularization can be thought of as some form of hashing, but the mapping function is injective. For worst-case volatility scenarios, optseq is empty, and the tag is always zero. For volatility shock scenarios, optseq = (τ, ξ, δ). It is straightforward to translate triples (τ, ξ, δ) into integer tags $2m$. Arranging lattice instances as shown in Fig. 9.3 and counting them from top to bottom, and from left to right, yields a valid sequence.

Tags are even-numbered. Odd-numbered tags are also used internally, reversing the evaluation view-point from sell-side to buy-side, or vice versa. Thus, if $\hat{F}(L)$ is computed on a lattice instance with regularized lattice signature $(\mathbf{X}, \lambda, 2m)$, then $-\hat{F}(L')$ is computed on lattice instance L' with signature $(\mathbf{X}, -\lambda, 2m+1)$. This reversal is necessary to compute the boundaries of corridors of uncertainty, namely $\hat{F}(L_n^U)$ and $-\hat{F}(L_n^D)$. (Recall that the signatures of L_n^U and L_n^D are $(X_n, 1)$ and $(X_n, -1)$, respectively. See Sect.8.2.1.)

With this information, the member elements of tScenario are as follows:

```
 1 class tScenario {
 2
 3 public:
 4
 5     enum tPosition {
 6         Buyer,
 7         Seller
 8     };
 9
10 private:
11
12     tPosition m_nPosition;
13
14 public:
15
16     virtual bool underControl( double gMultiplier );
17
18     virtual void refineExPolicy( tFDEngine& Engine,
19         int nBaseTag, int nIndex, double gDontExValue,
20         double gExValue, double gMultiplier,
21         tExPolicy& nExPolicy );
22
23     virtual double selectVol( int nTag, double gGamma,
24         double gMin, double gMax );
25
26     virtual bool endureOver( int nTag, double gNewTotal,
27         double gOldTotal );
28
29     virtual bool chooseOver( int nTag, double gNewTotal,
30         double gOldTotal );
31 };
```

Fig. 11.20. The definition of the abstract class `tScenario`. In general, worst-case scenarios are asymmetric for the buy- and sell-side. Which particular viewpoint is to be adopted is indicated by the value of `m_nPosition`

m_nPosition The nested type tPosition allows to flip the evaluation viewpoint *globally*. So far in this book, worst-case scenarios have always beed regarded from the seller's point of view, and prices have been maximized to cover the worst-case liability. To take the buyer's position, on the other hand, means to seek the smallest price to pay, in order to avoid loosing when the market behaves adversely. This changes the maximization to a minimization procedure: the "$\sup_{\sigma \in C}$" turns into an

11.5 Scenarios   169

"$\inf_{\sigma \in \mathcal{C}}$" in (4.10) and all the similar equations that follow. Simlarly, the "$\min_{A \subseteq A(L)} \max_{B \subseteq B(L)}$" first introduced in Def. 16 and occuring throughout Chapter 8 changes to "$\max_{A \subseteq A(L)} \min_{B \subseteq B(L)}$." This is because the interpretation of λ changes: positive λ now indicates a long position, whereas negative λ indicates a short position.

Any child class of tScenario is expected to initialize m_nPosition. For consistency with the earlier text, we assume here and below that m_nPosition = Seller.

underControl() If, for a given instrument X_n in the portfolio under consideration, $\lambda_n > 0$, then X_n is held short and not under the control of the agent. If $\lambda_n < 0$ the instrument is held long; potential early exercise is under control of the agent. underControl() interprets its parameter gMultiplier as λ_n and returns whether the corresponding position enables the agent to exercise control.

The situation is reversed if the global view-point changes from the sell-side to the buy-side.

refineExPolicy() The class tClaim uses the member function monitor() to produce an initial assessment of the local early exercise options for an individual instrument (see Sect. 11.1). If monitor() returns DontExercise or ForceExercise, the current lattice-instance node is assigned to the continuation respectively exercise region of the instrument for good. If monitor() returns MayExercise, as it does for standard American options, then the tScenario object is asked in turn to try to make a definite statement. Only when the tScenario object returns MayExercise as well is the current node assigned to the corridor of uncertainty of the instrument.

The safest policy is thus to return MayExercise throughout. However, as the tScenario object has access to other lattice instances through the Engine argument, a more advanced strategy such as described in Sects. 8.2.1 and 8.2.2 may be employed. This must be done in tScenario's child classes by overriding refineExPolicy().

The argument nBaseTag corresponds to the tag $2m$ of the regularized signature of the current lattice L. nIndex is the index of the claim in the portfolio, to be used as argument for tPortfolio::claim(). gDontExValue is the unit value of the instrument obtained through rollback. gExValue is the unit value of the instrument returned by tClaim::exercisePayoff() for the current node. gMultiplier is the number of contracts, and nExPolicy is an in-out parameter, initially set to MayExercise.

selectVol() The local volatility is selected between gMin and gMax. The actual implementation of the prototype in Fig. 11.20 may base its selection on gGamma, the finite difference approximation of $\frac{\partial^2}{\partial S^2} \hat{f}$, where $\hat{f}(S_t, t; L) = \hat{F}_t(L \mid S_t)$. The regularized tag of L is nTag. Two restrictions are immediately obvious:

- selectVol() works only for a one-factor model;
- selection schemes that require information beyond gamma cannot be realized.

The uncertain volatility models of Sects. 4.2.1, 4.2.2 and, with the introduction of a prior volatility parameter, 4.2.3, are all feasible, though.

endureOver() and chooseOver() This pair of functions is substituted for the max and min operators in the expression "$\min_{A \subseteq A(L)} \max_{B \subseteq B(L)}$" that occurs throughout Chapter 8 and in Figs. 8.4 and 8.6. The functions are folded over a sequence of values; gOldTotal represents the value selected thus far, and gNewTotal represents the new candidate. If the new value is to be selected over the old value, the function returns true.

Formula (8.95) of Sect 8.2.3 is used as a recipe for folding endureOver() over the arguments of the max operator and chooseOver() over the arguments of the min operator. The function names reflect the absence respectively presence of control by the agent.

11.5.2 Classes Derived from tScenario

tWorstCase is immediately derived from tScenario. Its definition is shown in Fig. 11.21. Figure 11.22 shows a listing of the member function selectVol(), and Fig. 11.23 shows the implementation of endureOver(). chooseOver() is implemented in an analogue fashion.

Figure 11.24 contains an outline of the function refineExPolicy(). The function consists of two branches, the first being executed if the pricing problem is linear or the corridors of uncertainty ought to be collapsed (in which case gDontExValue is the partial derivative of the worst-case value with respect to λ_{nIndex}). The second branch maintains the corridor of uncertainty by looking up the singleton portfolios $(X_{nIndex}, 1)$ and $(X_{nIndex}, -1)$. The function getClaim() of class tFDEngine does just that. This branch is an implementation of the algorithm in Fig. 8.7. Note that in addition to implementing the algorithm, refineExPolicy() must reverse all selection criteria if the global view-point is changed to the buy-side. This is done by setting the corrective constant nTag in line 17.

Only the case where gMultiplier is non-negative is shown in Fig. 11.24. The other case is handled symmetrically.

The class tShockScenario is an extension of tWorstCase. Figure 11.25 shows the definition. The interpretation of the member variables of tShockScenario follows Def. 22:

m_nDuration The duration parameter $d \geq 1$.
m_nPeriodicity The periodicity parameter $p \geq 1$.
m_nFrequency The frequency parameter $f \geq 1$

These variables are retrieved during rollback by a specialized compute engine of class tShockEngine. Since engines are passed only base references to

```
1  class tWorstCase : public tScenario {
2
3  public:
4
5     bool underControl( double gMultiplier );
6
7     void refineExPolicy( tFDEngine& Engine, int nBaseTag,
8         int nIndex, double gDontExValue, double gExValue,
9         double gMultiplier, tExPolicy& nExPolicy );
10
11    double selectVol( int nTag, double gGamma,
12        double gPrior, double gMin, double gMax );
13
14    bool endureOver( int nTag, double gNewTotal,
15        double gOldTotal );
16    bool chooseOver( int nTag, double gNewTotal,
17        double gOldTotal );
18 };
```

Fig. 11.21. tWorstCase instantiates the abstract member functions of tScenario

objects of class tScenario during initialization, a down-cast must be performed by tShockEngine to access these values. This is done safely with RTTI support.

Remark: the volatility shock scenario introduces additional events whose time points should be covered by the lattice. For that purpose, tScenario provides an (initially empty) method getEvGenerator() that is overridden by tShockScenario. This function is not shown in the figure.

11.6 Numerical Methods

MtgLib provides two methods to evaluate portfolios numerically: based on lattices, and with Monte Carlo simulation. Lattice-based evaluation is better supported at the time of this writing and the exclusive topic in the earlier parts of this book. For this reason we focus exclusively on the lattice-based facilities of MtgLib in the following paragraphs; see Chapter 13 for a brief description of Monte-Carlo extensions.

The classes that support lattice-based numerical evaluation fall into two categories: those that define the lattice template and manage lattice instances, and those that actually perform the finite difference rollback. Figure 11.26 shows both categories. The first group is capable of handling multi-factor models; the second group is not.

```
1  double tWorstCase::selectVol( int nTag, double gGamma,
2      double gMin, double gMax )
3
4  {
5      if( nTag % 2 == 0 ) {
6          switch( position() ) {
7              case Buyer :
8                  return ( gGamma <= 0 ) ? gMax : gMin;
9              case Seller :
10                 return ( gGamma >= 0 ) ? gMax : gMin;
11         }
12     }
13     switch( position() ) {
14         case Buyer :
15             return ( gGamma >= 0 ) ? gMax : gMin;
16         case Seller :
17             return ( gGamma <= 0 ) ? gMax : gMin;
18     }
19 }
```

Fig. 11.22. The body of the function tWorstCase::selectVol(). Depending on tag and global view-point, the function bases its decision on convexity respectively concavity. Recall that odd nTag indicates that $-\hat{F}(L')$ is being computed, where the signature of L' is $(\mathbf{X}, -\lambda, \text{nTag})$ and there exists a lattice instance L with signature $(\mathbf{X}, \lambda, \text{nTag} - 1)$. Since the negative signs are not actually applied, all comparisons need to be inverted

Some of the classes in Fig. 11.26 might as well be labeled "internal", since they are not directly visible through the scripting interface. They are listed here because of their proximity to the hierarchy of lattice-related classes, which is visible through the scripting interface.

11.6.1 Lattice Templates and Instances

The class tLattice describes the layout of the lattice. Number of factors, time discretization, space discretization, shape (tree or box) and space trimming determine the layout. Figure 11.27 shows how tLattice is defined.

The interpretation of the individual member components of tLattice is as follows:

m_pModel The lattice template needs to know about the model in order to create the entries for m_Space. For that purpose it uses the function tModel::createSpaceAxis().

11.6 Numerical Methods

```
1  bool tWorstCase::endureOver( int nTag, double gNewTotal,
2      double gOldTotal )
3
4  {
5      if( nTag % 2 == 0 ) {
6          switch( position() ) {
7              case Buyer :    // minimize
8                  return gNewTotal < gOldTotal;
9              case Seller :   // maximize
10                 return gNewTotal > gOldTotal;
11         }
12     }
13     switch( position() ) {
14         case Buyer :    // maximize
15             return gNewTotal > gOldTotal;
16         case Seller :   // minimize
17             return gNewTotal < gOldTotal;
18     }
19 }
```

Fig. 11.23. The body of the function tWorstCase::endureOver(). Depending on tag and global view-point, the function mimics a max or min operator

---

m_bIsBox The lattice can have the shape of a rectangular grid, or that of a tree, with the root labeled with S_0. This flag determines whether the rectangular grid shape is used.

m_bIsTrimmed and m_gTrimDev In order to reduce the running time, the lattice may be trimmed symmetrically at the outer regions. m_bIsTrimmed determines whether this is done. m_gTrimDev indicates the number of standard deviations after which the trimming should occur. The default values are true and 3.5. See [65] and the comment at the beginning of Sect. 5.1 for more details.

m_nMethod Can be either Explicit or Implicit and is used, among other things, as argument in calls to tModel::createSpaceAxis(), where it is used to ensure stability.

m_Bounds Determines the dimensions of the lattice layout when viewed as $(n+2)$-dimensional hypercube, where n is the number of factors, the $(n+1)$-st dimension is time and the last dimension is the combined vector of gradient and total portfolio value. m_Bounds is basically a sequence of pairs of upper and lower index bounds; the class tArrayBounds is not shown.

m_Space An array with one entry per factor. tSpaceAxis is an abstract class, because concrete implementations are model-dependent. Currently only

```
1  void tWorstCase::refineExPolicy( tFDEngine& Engine,
2      int nBaseTag, int nIndex, double gDontExValue,
3      double gExValue, double gMultiplier,
4      tExPolicy& nExPolicy )
5
6  {
7      double gValue;
8
9      if( Engine.isLinear() || Engine.accuracy() == Low ) {
10         if( gExValue > gDontExValue )
11             nExPolicy = xForceExercise;
12         else
13             nExPolicy = xDontExercise;
14     }
15     else {
16         if( gMultiplier >= 0 ) {
17             int nTag = ( position() == Buyer ) ? 0 : 1;
18
19             if( gExValue > gDontExValue ) {
20                 Engine.getClaim( nIndex,
21                     nBaseTag + 1 - nTag, gValue );
22                 if( gExValue > gValue )
23                     nExPolicy = xForceExercise;
24             }
25             else {
26                 Engine.getClaim( nIndex,
27                     nBaseTag + nTag, gValue );
28                 if( gExValue <= gValue )
29                     nExPolicy = xDontExercise;
30             }
31         }
32         else {
33             // the other case is symmetric
34         }
35     }
36 }
```

Fig. 11.24. An outline of the function refineExPolicy() of class tWorstCase. The constant Low in line 9 corresponds to the strategy to collapse corridors of uncertainty; the else branch maintains corridors of uncertainty

```
1  class tShockScenario : public tWorstCase {
2
3      int m_nDuration;
4      int m_nPeriodicity;
5      int m_nFrequency;
6  };
```

Fig. 11.25. The class tShockScenario merely adds the parameters of a volatility shock scenario as defined in Def. 22

Class name	Purpose
tLattice	Lattice template
tTimeAxis	Discretization of time
tSpaceAxis	Discretization of space for one factor (abstract)
tGeoSpaceAxis	Discretization of space based on geometric Brownian motion
tLatticeInstance	Lattice instance, what else?
tOFSolver	One-factor finite difference solver (abstract)
tOFExplicit	Explicit finite difference solver
tGeoExplicit	Explicit solver for models based on geometric Brownian motion
tOFImplicit	Crank-Nicholson finite difference solver
tGeoImplicit	Crank-Nicholson solver for models based on geometric Brownian motion
tGeoSolver	Additional base class for tGeoExplicit and tGeoImplicit (abstract)

Fig. 11.26. The collection of classes that work together to support lattice-based evaluation. The prefix "OF" stands for one-factor. tGeoExplicit and tGeoImplicit have two parent classes and are thus a case of multiple inheritance

    tGeoSpaceAxis for the model class tBSModel is implemented. Note that the dimension of each space axis must be consistent with the corresponding information in m_Bounds.

m_pTime The discretization of the time axis, which may be non-uniform. The time axis is only finalized after all the space axes have been created, for the required cap on the size of the largest time step can only then be known (see output dt of the algorithm in Fig. 5.4). The class tTimeAxis is final. tTimeAxis is purely mathematical and does not support real calendar dates.

createOFSolver() Finite difference solvers are discussed in the next section. This function creates a solver for one-factor models. Its implementation is simple: since the finite difference approximation of partial derivatives

```
 1 class tLattice {
 2
 3     tModel* m_pModel;
 4
 5     bool m_bIsBox;
 6     bool m_bIsTrimmed;
 7     double m_gTrimDev;
 8
 9     tFDMethod m_nMethod;
10
11     tArrayBounds m_Bounds;
12
13     tSpaceAxis* m_Space[...];
14     tTimeAxis* m_pTime;
15
16 public:
17
18     tOFSolver* createOFSolver();
19
20     tRetCode createInstance( tPortfolio* pPf,
21         const tSignature* pSig, int nTag,
22         tLatticeInstance*& pInstance ) const;
23 };
```

Fig. 11.27. The class tLattice defines the layout of the lattice (the lattice template), from which lattice instances are created by calling createInstance()

depends on the geometry of the discretization as well as the underlying stochastic process, the request is forwarded to the space axis, which knows about these properties:

```
1 tOFSolver* tLattice::createOFSolver()
2
3 {
4     return m_Space[0]->createOFSolver();
5 }
```

Multi-factor solvers are not implemented.

createInstance() The lattice template is also used to create lattice instances of it. The signature of the new lattice instance is implied by the arguments pPf, pSig and nTag. tSignature is implemented as a bitfield; its precise definition is not shown.

```cpp
class tLatticeInstance {

    tClaim* m_Slot[...];

    int m_nCurrent;

    tMultiArray<double> m_Buffer[2];

    tMultiArray<double> m_Prep;
    tMultiArray<double> m_Temp1;
    tMultiArray<double> m_Temp2;

public:

    void beforeRollback( int nDay );
    void afterRollback( int nDay );

    void rotate();

    tMultiArray<double>& current() {
        return m_Buffer[m_nCurrent];
    }

    tMultiArray<double>& last() {
        return m_Buffer[m_nLast];
    }
};
```

Fig. 11.28. A *very* condensed summary of class tLatticeInstance. The template tMultiArray allows arrays whose dimensions are determined by objects of class tArrayBounds. m_Prep, m_Temp1 and m_Temp2 can, but must not be used during rollback

The lattice instances created by createInstance() in tLattice belong to the class tLatticeInstance, a very condensed definition of which is shown in Fig. 11.28.

Lattice instances do not allocate memory for the entire grid, but only for two adjacent hyperplanes, cut perpendicular to the time axis. This is standard procedure for memory-aware implementations of one-level finite-difference schemes and tree methods. One hyperplane contains the values for the previously processed time slice t_{i+1}, the other receives the result of the current rollback round for time slice t_i. (In Sect. 8.2.3 we have seen that this can lead to considerable slowdown due to restart.)

Additional temporary space may be necessary. In Sect. 8.2.3, a scheme to save intermediate results of the minmax calculation has been proposed to increase the efficiency slightly. m_Prep is used for this purpose. Also, some finite difference solvers may need their own scratch space; Crank-Nicholson, for instance, requires extra storage for the decomposed coefficient matrix and the right-side vector of the linear system which it has to solve. m_Temp1 and m_Temp2 can be activated for that purpose.

In summary, the components of tLatticeInstance shown in Fig. 11.28 have the following meaning:

m_Slot The portfolio and signature arguments passed to createInstance() in tLattice are converted into a compact array m_Slot of references to instruments.

The definition of m_Slot as array of references to instruments is incomplete, however. Instruments have different maturity dates and thus enter into the computation at different times during the rollback, therefore widening the lattice instance dynamically (of course, all memory is allocated before-hand, and the widening is only logical). m_Slot has additional features to allow this process to occur efficiently. There are also some additionl supporting members, for instance for index translation between m_Slot and tPortfolio::m_Claim. All this is not shown for simplicity.

m_nCurrent The index in m_Buffer of the multi-array representing the current hyperplane. "Current" refers to the time slice, say t_i, that is being computed in the current rollback round. The "last" hyperplane refers to the hyperplane of time slice t_{i+1}.

m_Buffer This buffer holds two hyperplanes of the total space of the lattice instance. m_Buffer[m_nCurrent] contains the current hyperplane; m_Buffer[m_nCurrent − 1] contains the last hyperplane. We stress again that the innermost coordinate of each hyperplane loops through the gradient $\hat{v}_k$ plus the total worst-cast value $\hat{V}$.

m_Prep Temporary space used for intermediate results (see Sect. 8.2.3).

m_Temp1 and m_Temp2 Temporary space, mainly used by mixed explicit/implicit schemes such as Crank-Nicholson.

beforeRollback() and afterRollback() These functions are called before and after rollback rounds when the current time slice t_i falls on a day boundary. These functions take care of maturing instruments and adjust the (logical) width of the lattice instance.

rotate() Replaces m_nCurrent with m_nCurrent − 1 and thus rotates the current and last hyperplanes.

current() Returns a reference to the current hyperplane.

last() Returns a reference to the last hyperplane.

current() and last() are not the only functions to access the elements of the lattice instance. Their are additional functions to read and write m_Prep, m_Temp1 and m_Temp2. There are also functions to access not entire multi-arrays, but single innermost rows or entries. The list of actual member functions exceeds the list of functions shown in Fig. 11.28.

The classes tTimeAxis and tSpaceAxis are less interesting. We only note that tSpaceAxis contains the virtual member createOFSolver() mentioned above, and that the derived class tGeoSpaceAxis implements this function as follows:

```
1  tOFSolver* tGeoSpaceAxis::createOFSolver()
2
3  {
4      if( isImplicit() )
5          return new tGeoImplicit( this );
6      return new tGeoExplicit( this );
7  }
```

tGeoSpaceAxis also implements the algorithm in Fig. 5.4 to find a stable discretization.

11.6.2 Finite Difference Solvers

Figure 11.26 shows the hierarchy of finite difference solvers, but does not emphasize the multiple-inheritance relationship very strongly. This is done in Fig. 11.29, which shows the dependency graph. Doubly framed classes are abstract base classes which are ultimately used as interfaces to access the functionality of the concrete solver.

The purpose of each class is as follows:

tOFSolver This base class is general in the sense that its member functions rely on the assembly of the tridiagonal coefficient matrix and the right-side vector for one rollback round at some other place. (Both explicit and mixed explicit/implicit methods can be expressed in this manner.) Once the linear system of equations has been set up, its solution can be computed independently from the concrete financial model or spatial lattice geometry. The most visible feature of tOFSolver is the pure virtual member function solve().

tOFExplicit Provides a body for the prototype solve() in tOFSolver. Uses an explicit forward Euler one-level scheme.

tOFImplicit Provides a body for the prototype solve() in tOFSolver. Uses a mixed explicit/implicit Crank-Nicholson scheme. In addition, allows incremental refinement, which is necessary for American options.

tGeoSolver This abstract class contains a reference to the tGeoSpaceAxis object that has created the solver. It also has access to the model drift and volatility coefficients from which to build the tridiagonal transition matrix. tGeoSolver acts as pheripheral source of information.

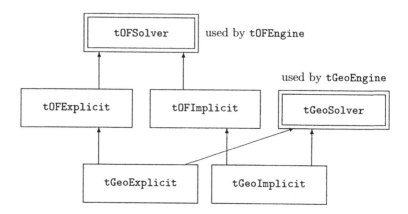

Fig. 11.29. The class hierarchy for finite difference solvers. The abstract base class tOFSolver is used by member functions of tOFEngine to access the functionality of a particular solver. Similarly, the abstract base class tGeoSolver is used by tGeoEngine

tGeoExplicit This class bridges between tGeoSolver and tOFExplicit. If retrieves model coefficients form the former and instantiates the transition weights for the latter. The simplified nature of the linear system of equations in the explicit case is taken into account.

tGeoImplicit This class bridges between tGeoSolver and tOFImplicit. If retrieves model coefficients form the former and instantiates the transition matrix and right-side vector for the latter.

tOFSolver and Child Classes Some of the essential features of tOFSolver are shown in Fig. 11.30. A description follows:

tProcessParamsStub This empty type provides a handle to whatever process parameters need to be transferred in order to compute the transition matrix. This type is public. In our case, it is expanded by tGeoSolver to provide the local drift and volatility coefficients.

tIncrement If American options are involved and a mixed explicit/implicit scheme is used, the solution of the linear system of equations needs to be refined. Iterative refinement proceeds in alternatingly re-evaluating early-exercise decisions and performing a subsequent over-relaxation step. This loop may require additional memory space or knowledge, which is encapsulated in a class derived from tIncrement. The iterative refinement loop is then executed by calling the virtual member functions of tIncrement.

calcWeights() This function must compute the transition weights. The arguments nFromLevel and nToLevel indicate the location of the interval

```
1  class tOFSolver {
2
3  public:
4
5      struct tProcessParamsStub {
6      };
7
8      class tIncrement {
9          public:
10             virtual void beginIncrement( ... );
11             virtual void doIncrement( ... );
12             virtual void endIncrement( ... );
13     };
14
15 protected:
16
17     virtual void calcWeights( int nFromLevel, int nToLevel,
18         const tProcessParamsStub& Params );
19
20 public:
21
22     virtual void solve();
23     virtual tRetCode refine( tIncrement& Incr );
24 };
```

Fig. 11.30. Some of the features of tOFSolver. The virtual functions *and* the local types must be expanded in child or otherwise related classes

of nodes in the lattice for which the rollback is being performed (the interval may change due to knock-out or a tree-shaped lattice). The function also gets to see the process parameters in the argument Params, after a proper down-cast.

solve() The top-level function that initiates the current rollback round. This function contains the numeric part of the code.

refine() If solve() has successfully finished and the portfolio contains American options, refine() must be called repeatedly to adjust the result. This function makes use of the tIncrement interface.

tOFExplicit is a straightforward instantiation of tOFSolver. The function calcWeights() remains still unresolved, since tOFExplicit is a general explicit solver that does not know about the concrete model coefficients. solve() is implemented, but the function refine() is ignored.

The class tOFImplicit is more complex. It implements both solve() and refine(). solve() performs a LU decomposition based on the LAPACK modules DGTTRF and DGTTRFS, which were translated and adapted from Fortran. The code allows for partial pivoting. See [61] for more details.

The function refine() reuses the decomposition of solve() to modify the current result. It does so in an over-relaxation step where the relaxation parameter ω lies between 1 and 2 and is dynamically adapted, based on the previous iteration count. Over-relaxation is explained in [79].

tGeoSolver and Child Classes The auxiliary class tGeoSolver is defined in Fig. 11.31. It mainly expands the empty type tProcessParamsStub defined in class tOFSolver. The elements m_gVol, m_gDrift and m_gDiscount are the local values of the model coefficients in class tBSModel. The process parameters are retrieved from the model by the compute engine, which is an instance of type tGeoEngine (more on compute engines below).

```
 1  class tGeoSolver {
 2
 3  public:
 4
 5      struct tProcessParams :
 6              public tOFSolver::tProcessParamsStub {
 7          double m_gVol;
 8          double m_gDrift;
 9          double m_gDiscount;
10      };
11  };
```

Fig. 11.31. The class tGeoSolver expands the stub class tProcessParamsStub, defined in tOFSolver. Objects of class tProcessParams will be supplied by the compute engine

The classes tGeoExplicit and tGeoImplicit combine the model specific information captured in tGeoSolver and the numerical functionality of tOFExplicit and tOFImplicit. Currently only these two types of solvers are implemented in MtgLib. The definition is shown in Fig. 11.32. The implementation of calcWeights() is omitted.

11.7 Evaluators

Evaluators collect portfolio, model, scenario and numerical method, make sure they all fit together and initiate the evaluation process. Evaluators are short-lived, contrary to objects of other external classes which can be reused

```
1  class tGeoExplicit : public tOFExplicit, public tGeoSolver {
2
3      void calcWeights( int nFromLevel, int nToLevel,
4          const tProcessParamsStub& Params );
5  };
6
7  class tGeoImplicit : public tOFImplicit, public tGeoSolver {
8
9      void calcWeights( int nFromLevel, int nToLevel,
10         const tProcessParamsStub& Params );
11 };
```

Fig. 11.32. Actual solvers belong either to class tGeoExplicit or tGeoImplicit. They compute the transition matrix from information provided in Params

(note that the evaluator object in Fig. 11.1 does not need a name). They are used once and then thrown away. Figure 11.33 shows the definition.

```
1  class tEvaluate {
2
3      tPortfolio* m_pPortfolio;
4      tModel* m_pModel;
5      tScenario* m_pScenario;
6      tOptimizer* m_pOptimizer;
7      tLattice* m_pLattice;
8      tPathSpace* m_pPathSpace;
9
10     tFDEngine* m_pFDEngine;
11     tMCEngine* m_pMCEngine;
12
13     tCurveContainer m_CurveContainer;
14     tImageContainer m_ImageContainer;
15
16     tAccuracy m_nAccuracy;
17
18 public:
19
20     tRetCode run();
21 };
```

Fig. 11.33. Evaluators know about all the objects that make up a particular pricing problem, cross-reference them and oversee the evaluation process

11 The Class Hierarchy of MtgLib—External

The individual components of tEvaluate have the following semantics:

m_pPortfolio A reference to the portfolio to be evaluated.

m_pModel A reference to the model under which the portfolio is to be evaluated.

m_pScenario A reference to the scenario for the evaluation. This parameter is ignored if Monte-Carlo is the numerical method of choice.

m_pOptimizer At the time of this writing, calibration through optimization is only possible for Monte-Carlo methods. The optimizer object adds an outer loop to the evaluation process, which the compute engine must know about. Ignored if lattice-based evaluation is selected.

m_pLattice If this pointer is set, the evaluation is lattice-based.

m_pPathSpace If this pointer is set, the evaluation is done with Monte-Carlo simulation. Only one of m_pLattice and m_pPathSpace can be set. The class tPathSpace is not explained here.

m_pFDEngine The compute engine created with the call

    m_pModel->createEngine(m_pScenario, m_pFDEngine,
        m_nAccuracy);

if m_pLattice is set. See also Fig. 11.8.

m_pMCEngine The compute engine created with the call

    m_pModel->createEngine(m_pMCEngine);

if m_pPathSpace is set.

m_CurveContainer Besides pricing and optimization (under Monte-Carlo), MtgLib also offers curve-generating functionality from calibrated path spaces. Curves can be written to files, to be used in subsequent pricing rounds.

m_ImageContainer Calibrated curves can also be converted into images. Currently supported is the PNG format popular on the Internet. Curves and images are produced by the Monte-Carlo extensions described in Chapter 13.

m_nAccuracy This parameter applies to lattice-based evaluation only and controls which of the speed-up techniques of Sect. 8.2 should be used. Possible values are Low (collapsing the corridors of uncertainty for American options) and Exact (maintaining the corridors of uncertainty). See also Sect. 11.3.

run() This functions initiates the evaluation process. It transfers control to the run() member function of either *m_pFDEngine or *m_pMCEngine.

12 The Class Hierarchy of MtgLib—Internal

Most classes in MtgLib are internal—not visible through the scripting language, of which an example is given in Fig. 11.1. In this section, we discuss the most important category of internal classes: compute engines.

12.1 Compute Engines

Figure 12.1 shows the current hierarchy of compute engines. tFDEngine and its subclasses are used for lattice based evaluation. tMCEngine and its subclasses are used for Monte-Carlo simulation. We only discuss the branch emanating from tFDEngine; see Chapter 13 for some remarks on Monte-Carlo.

Class name	Purpose
tEngine	Base class (abstract)
tFDEngine	Extended base class for lattice-based evaluation (abstract)
tOFEngine	Extended base class for one-factor lattice-based evaluation (abstract)
tGeoEngine	Worst-case evaluation for geometric Brownian motion models
tShockEngine	Evaluation under volatility shock scenarios
tMCEngine	Extended base class for Monte-Carlo simulation (abstract)
tHJMEngine	For HJM interest-rate models
tShortRateEngine	For the Vasicek model, for example

Fig. 12.1. The hierarchy of compute engines. tShockEngine extends tGeoEngine by overriding some functions that handle administrative tasks between rollback rounds. Fixed income classes are not discussed here

tEngine, tFDEngine and tOFEngine are all abstract; they do not function by themselves. Instances are created from tGeoEngine and tShockEngine, depending on which scenario is used (see Fig. 11.11). The classes are used as follows:

tEngine This class has several purposes:
- It contains references to objects used by all types of compute engines, like the model or portfolio.
- It performs some initialization that can be done on that high a level; for instance, it matches factors referenced in the portfolio with factors defined in the model.
- It defines an interface to retrieve singleton portfolios. This is important for corridors of uncertainty.

Many functions that need to access the current state (the payoff functions in Fig. 11.3, for instance) do so either through the tEngine or through the tFDEngine interface, after a proper down-cast.

tFDEngine This class extends tEngine in several regards:
- It contains more references to objects defined for the current problem, for instance the lattice template and the scenario.
- It implements the main run() function of any compute engine derived from it. tFDEngine contains local data structures for the dynamic lookup mechanism of lattice instances through regularized signatures. In some sense, tFDEngine implements the "outer loop" over the time axis of the finite difference scheme.
- It contains more information about the current state than tEngine. It knows which time slice t_i is currently being processed (since it drives the loop!), and provides variables, to be set by subclasses, that locate individual nodes in the corresponding hyperplane. It provides functions to access this state information.

tFDEngine is equipped to handle multi-factor models. This class implements the core of the multi-lattice dynamic programming paradigm introduced in Sect. 5.1.

tOFEngine tFDEngine is compatible with multi-factor models; tOFEngine reduces the number of supported factors to one. tOFEngine adds inner-loop functionality to tFDEngine. (For the inner loop for fixed t_i, the number of factors must be known.)

tGeoEngine This class instantiates tOFEngine to support one-factor geometric Brownian motion models under worst-case volatility scenarios.

tShockEngine This class extends tGeoEngine to support volatility shock scenarios.

The following paragraphs go into some implementation details.

12.1.1 The Abstract Class tEngine

A shortened definition of tEngine is shown in Fig. 12.2. The meaning of the individual member components is as follows:

```
1  class tEngine {
2
3      tPortfolio* m_pPortfolio;
4      tModel* m_pModel;
5
6      int m_FactorXlat[...];
7
8  protected:
9
10     virtual tRetCode beforeRun();
11     virtual tRetCode afterRun();
12
13 public:
14
15     virtual void getClaim( int nIndex, int nTag,
16          double& gUnitValue );
17 };
```

Fig. 12.2. The abstract base class tEngine performs some preparation and cleanup tasks before and after evaluation. It also defines the interface for the retrieval of singleton portfolios, which is important for finding the corridors of uncertainty for American options

m_pPortfolio A reference to the portfolio object under investigation.

m_pModel A reference to the model used for the evaluation.

m_FactorXlat The member function matchFactors() of class tPortfolio unifies the factor tables in the portfolio and model objects (see Fig. 11.6). The resulting index permutation is stored in m_FactorXlat.

beforeRun() This function takes care of initialization issues that can be handled with limited information. Factor matching is an example. If this function is overridden in child class tXYZEngine, the original function must always be called first:

```
1  tRetCode tXYZEngine::beforeRun() {
2      if( tEngine::beforeRun() != OK ) ...
3      ...
4      return OK;
5  }
```

As the class hierarchy builds up, each class contributes to initialization by overriding beforeRun(), but still calling the function in the parent class.

afterRun() Performs cleanup jobs, mostly related to memory management, after evaluation has been completed. Again, overriding functions must make sure to call the original eventually:

```
1  tRetCode tXYZEngine::afterRun() {
2    ...
3    return tEngine::afterRun();
4  }
```

getClaim() This function is a pure virtual interface to retrieve the current value of the singleton portfolio with signature $(X_{\text{nIndex}}, \alpha)$, where $\alpha = \pm 1$, depending on nTag and scenario settings such as sell-side or buy-side point-of-view. The function is called in tWorstCase::refineExPolicy(), for instance, as shown in Fig. 11.24.

12.1.2 The Abstract Class tFDEngine

tFDEngine implements the loop over the time axis and handles the repository of lattice instances. It works independently of the number of factors. Figure 12.3 shows the relevant fragment of its definition. Individual members are used as follows:

m_nDay and m_gFractionOfDay The value of t_i in days, where t_i is the current time slice. More precisely,

$$t_i = \text{m\_nDay} + \text{m\_gFractionOfDay}$$

The distinction between days and fractions of days is convenient, because the granularity for events connected with instruments or scenarios is at the level of days.

m_pLattice A reference to the lattice template.

m_pScenario A reference to the scenario object.

m_Pos This variable is not maintained by tFDEngine, only provided in order to be accessible through it. Any subclass that loops over the nodes of the current time slice t_i must update m_Pos during the preparatory phase of each loop iteration. The preparatory phase ends once the finite difference solver takes over.

To keep m_Pos consistent is important: functions like tClaim::payoff() must know exactly which node is currently being processed, because they are called node for node.

doRound() Executes exactly one rollback round for all currently known lattice instances. This function is final; it calls the function doTask(), which must be instantiated in any subclass, to process each lattice instance. doRound() observes the rule proposed in 5.1.1 for external consistency, augmented by provisions that guarantee that lattice instances are processed in the correct order under volatility shock scenarios (in fact, any scenario that uses a consistent pattern for regularizing signatures).

doTask() The pure virtual prototype that is called by doRound().

getLatticeInstance() Accesses the lattice instance whose regularized signature is determined from the pair Sig/nTag. If the lattice instance does

```
1  class tFDEngine : public tEngine {
2
3      int m_nDay;
4      double m_gFractionOfDay;
5
6      tLattice* m_pLattice;
7      tScenario* m_pScenario;
8
9      int m_Pos[...];
10
11     tRetCode doRound();
12
13 protected:
14
15     virtual void doTask( tLatticeInstance& Instance );
16
17     void getLatticeInstance( const tSignature& Sig,
18         int nTag, tLatticeInstance*& pInstance );
19
20     tRetCode beforeRun();
21     tRetCode afterRun();
22
23 public:
24
25     tRetCode run();
26
27     void getClaim( int nIndex, int nTag,
28         double& gUnitValue );
29 };
```

Fig. 12.3. A small part of the definition of tFDEngine. Shown are the state information, the interface to access lattice instances, and the main run() function

not exist, getLatticeInstance() creates it and interrupts the current iteration of doRound() through the C++ exception mechanism.

beforeRun() After calling its parent function, beforeRun() initializes the repository of lattice instances and creates the top-level instance.

afterRun() Merely calls the parent functions, and does not do any additional processing.

run() This function contains the central control loop over the time domain, as shown in Fig. 12.4.

getClaim() Instantiates the virtual function getClaim() whose prototype is defined in Fig. 12.2.

```
 1  tRetCode tFDEngine::run()
 2
 3  {
 4      if( ( nRet = beforeRun() ) != OK )
 5          return nRet;
 6
 7      int nNumOfRounds = m_pLattice->numOfSlices();
 8
 9      for( int k = 0; k < nNumOfRounds; ++k ) {
10          if( ( nRet = doRound() ) != OK ) {
11              cleanup();
12              return nRet;
13          }
14      }
15
16      return afterRun();
17  }
```

Fig. 12.4. A schematized listing of the central control loop in the function run() of class tFDEngine. numOfSlices() is not included in Fig. 11.27; it returns the number of discretization points in the time domain. doRound() belongs to tFDEngine and executes one rollback round

12.1.3 The Abstract Class tOFEngine

tOFEngine is the last abstract class in the chain of ever more specialized classes for compute engines.

tOFEngine provides the inner-loop functionality for one-factor models. The member function doTask() performs one rollback-round for the lattice instance passed to it as argument. What is missing to make tOFEngine a full-fledged compute engine is the calculation of the local model coefficients for the solver.

Figure 12.5 shows the definition of class tOFEngine. The members of tOFEngine have the following semantics:

tIncrement The empty interface tIncrement has been defined in Fig. 11.30 for class tOFSolver to support incremental refinement for mixed explicit/implicit schemes. Here, the interface is instantiated as a proxy that forwards all requests to the parent compute engine. nAdjUp and nAdjDown are the adjusted number of nodes above and below the centered root node of the lattice. Adjustments occur when instruments knock out and therefore set up a new boundary.

```
1  class tOFEngine : public tFDEngine {
2
3      class tIncrement : public tOFSolver::tIncrement {
4          tOFEngine& m_Engine;
5
6          tIncrement( tOFEngine& Engine )
7              : m_Engine( Engine ) {}
8
9          void beginIncrement( int nAdjDown, int nAdjUp ) {
10             m_Engine.beginIncrement( nAdjDown, nAdjUp ); }
11         void doIncrement( const int Pos[...] ) {
12             m_Engine.doIncrement( Pos );            }
13         void endIncrement( int nAdjDown, int nAdjUp ) {
14             m_Engine.endIncrement( nAdjDown, nAdjUp ); }
15     };
16
17     tOFSolver* m_pSolver;
18
19     void doBarriers( int& nAdjDown, int& nAdjUp );
20     void doBoundary( int nAdjDown, int nAdjUp );
21     void doRollback( int nAdjDown, int nAdjUp );
22     void doMonitor( int nAdjDown, int nAdjUp );
23     void doPayoff();
24
25     void beginIncrement( int nAdjDown, int nAdjUp );
26     void doIncrement( const int Pos[...] );
27     void endIncrement( int nAdjDown, int nAdjUp );
28
29 protected:
30
31     virtual const tOFSolver::tProcessParamsStub&
32         getProcessParams();
33     virtual tRetCode createSolver( tOFSolver*& pSolver );
34
35     void doTask( tLatticeInstance& Instance );
36
37     tRetCode beforeRun();
38     tRetCode afterRun();
39 };
```

Fig. 12.5. The class tOFEngine: a compute engine for one-factor models

12 The Class Hierarchy of MtgLib—Internal

m_pSolver The finite difference solver created with a call to createSolver(), see below.

doBarriers() The rollback round is executed in stages. Each stage is dedicated to a sub-task. doBarriers() locates all the barriers and initializes temporary data structures that guide the subsequent sub-tasks. It also returns the location of the adjusted boundary in nAdjUp and nAdjDown. These values are used in all subsequent sub-tasks.

doBoundary() Performs the second sub-task. If barriers have been found, lattice instances for subordinate partial portfolios must be accessed to set the boundary data. This is done by calling getLatticeInstance().

doRollback() Once the boundary has been taken care of, the "numerical" part of the rollback (i.e., what is commonly associated with the term) is done for the continuation region. After some preparation, this function essentially calls m_pSolver->solve().

doMonitor() If American options are present, early exercise policies are gathered for each node by calling tClaim::monitor() for all relevant instruments and refining estimates with m_pScenario->refineExPolicy(). Then, early exercise combinations are evaluated where alternatives exist. This is the minmax calculation, with exploitation of intermediate results as outlined in Sect. 8.2.3.

beginIncrement(), doIncrement(), endIncrement() These functions repeat the monitoring of American options, given the current result. They contribute to the incremental refinement in the over-relaxation method used under mixed explict/implicit schemes. Refinement is started with m_pSolver->refine() before doPayoff() is called.

doPayoff() In the final sub-task, payoffs of maturing instruments and fixed cashflows are added. doPayoff() is only called *after* incremental refinement through the tIncrement proxy has been completed.

getProcessParams() This pure virtual function must be instantiated by a subclass. It supplies the missing information on which evaluation relies. To defer the instantiation of getProcessParams() makes tOFEngine general.

createSolver() This function creates the finite difference solver: it calls m_pLattice->createOFSolver(). (The lattice template, in turn, relays the request to the space axis, as described in Sect. 11.6.1.)

doTask() The main function of the class calls doBarriers(), doBoundary(), doRollback(), doMonitor(), refines, and calls doPayoff(), in that order.

beforeRun() After calling the parent version, this function creates the solver by calling createSolver().

afterRun() Deletes the solver.

tFDEngine and tOFEngine together are the logistic heart of lattice-based evaluation (the solver and lattice class hierarchies are the numerical one), comprising combined about 2800 lines of code.

12.1.4 The Concrete Class tGeoEngine

Not much remains to do to complete tOFEngine to a working compute engine for one-factor geometric Brownian motion models. Figure 12.6 shows the rather short definition.

```
1 class tGeoEngine : public tOFEngine {
2
3     tGeoSolver::tProcessParams m_Params;
4     const tOFSolver::tProcessParamsStub& getProcessParams();
5 };
```

Fig. 12.6. The class tGeoEngine prepares the model coefficients for tOFEngine and tGeoSolver

Class tGeoEngine ensures that the solver receives the correct model coefficients for the current time slice. The function getProcessParams() reads the drift and volatility bounds from the model (to which a reference is provided in tEngine). It computes the local gamma and uses the scenario object (to which a reference is kept in tFDEngine) to select the scenario volatility, by calling the member function selectVol(). Process parameters are then stored in m_Params and returned.

12.1.5 The Concrete Class tShockEngine

```
1 class tShockEngine : public tGeoEngine {
2
3     int m_nDepth;
4
5     int m_nCurTag;
6     tLatticeInstance* m_pCurCoInstance;
7
8     tRetCode beforeRun();
9 };
```

Fig. 12.7. The class tShockEngine and some of its members

The class tShockEngine extends tGeoEngine for volatility shock scenarios, which require periodic data transfers between lattice instances. Figure 12.7 shows some of the members of tShockEngine. They are interpreted as follows:

m_nDepth The number of conventional lattice instances per consolidating lattice instance. If d is the duration of the volatility shock scenario, and p its periodicity, then m_nDepth $= \lceil d/p \rceil$. See Sect. 9.1.1 for motivation.

m_nCurTag The regularized tag of the current lattice instance. m_nCurTag and the scenario parameters d, p and f imply the extended signature variables τ, ξ and δ as defined in Def. 23.

m_pCurCoInstance If the extended-signature parameter τ of the current lattice instance is "conventional", then m_pCurCoInstance references the consolidating lattice instance of the same level. If the current lattice instance is consolidating, then m_pCurCoInstance points to the conventional lattice instance from which data might have to be imported. (Data is only compared and possibly imported on certain dates, and only with respect to one conventional lattice instance at a time.)

beforeRun() Creates all the extra lattice instances needed for the volatility shock scenario. This function implements the algorithm of Fig. 9.4.

12.2 Other Groups of Classes

MtgLib contains about 135 classes, of which only those that form the combinatorial and mathematical kernel have been discussed in the previous sections. Other classes fill in the infrastructure to create actual applications.

These categories of classes are also part of MtgLib:

- Figure 11.1 shows an example of the scripting language in which MtgSvr communicates. In general, each object class knows how to parse itself. Each object class contains a static member function parse() that creates a new object from a script definition. There is a central class tParser, and peripheral classes tScanner, tSource, tFileSource, tStringSource and tNetSource for support. For customized claims, classes tExpression, tNumericalExpr and tExPolicyExpr provide the necessary extension to the scripting language. The parser is of the recursive-descent variety.
- MtgLib contains many classes for bond mathematics: for example, tBondEquations, tBondMath, and tUSBondMath handle yield calculations.
- MtgSvr resides as a service on Windows NT PC's. The classes tSocket, tService, tJobServer support this background operation.
- Some low-level classes provide special data structures: tHeap and tHeap2 for one- and two-dimensional dynamic arrays; tMultiArray; tMap for one-dimensional, highly homogeneous arrays; and tSignature, which is implemented as a bitfield.

13 Extensions for Monte-Carlo Pricing and Calibration

Faithful to the theme of the book, the previous sections 11 and 12 have focused on scenario-based pricing of portfolios of vanilla, barrier and American options for equity and FX Black-Scholes models.

In this section we outline some extensions for minimum-entropy pricing and calibration, which were described in Sect. 4.3.

Monte-Carlo compute engines for Monte Carlo have already been listed in Fig. 12.1. Classes for interest rate models are listed in Fig. 11.7.

Figure 13.1 shows additional classes for the outer optimization loop (recall that tEvaluate in Fig. 11.33 contains a member variable m_pOptimizer).

Class name	Purpose
tOptimizer	optimizer template (abstract)
tEntropyOpt	minimum entropy optimizer template
tOptInstance	optimizer instance (abstract)
tMCOptInstance	optimizer instance for Monte Carlo (abstract)
tMCEntropyOptInstance	optimizer instance for minimum entropy optimization (also inherits from tMinimizer)
tMinimizer	wrapper for L-BFGS-B

Fig. 13.1. Extensions to MtgLib. Not shown in this picture are the compute engines derived from tMCEngine; these are mentioned in Sect. 12.1

The purpose of each class in Fig. 13.1 is as follows:

tOptimizer The abstract base class for optimizer templates. Just as for lattices, we distinguish between optimizer templates that contain information on the type of the optimizer, and actual optimizer instances which are created by optimizer templates as requested by compute engines and used only once. Optimizer instances contain the "dirty" variables used during the actual computation.

The main member function of tOptimizer is the createInstance(), which is pure virtual.

tEntropyOpt Supports minimum entropy optimizer templates. The function createInstance() returns objects of type tMCEntropyOptInstance. In addition, minimum entropy optimizer templates specify the upper and lower bound for Lagrange multipliers. The following fragment would be a valid definition of a minimum entropy optimizer in the scripting language MtgScript:

```
1 optimizer xyz {
2     type entropy,
3     low -100, high 100
4 }
```

tOptInstance The abstract base class for optimizer instances. This class is designed with both lattice-based and simulation methods in mind, although optimization right now is supported only for simulation methods. tOptInstance contains basic member variables such as m_Price (the price vector $\bar{\pi}$), m_Lambda (the output vector holding the optimal $\hat{\lambda}$'s) and m_Gradient (used by the gradient-based minimization routine).

tMCOptInstance A specialization of tOptInstance for Monte Carlo simulation. The member variable m_Weight is the vector that holds the alternative distribution Q for the Monte Carlo paths. This class also precomputes the discounted cashflows for each instrument and path, since this information needs to be computed only once. It then calls a pure virtual member function minimize() to do the actual optimization.

tMCEntropyOptInstance Inherits from tMCOptInstance and tMinimizer; implements minimize() defined in tMCOptInstance by passing control to tMinimizer::minimize(), which in turn calls back a member function eval() in each iteration. eval() is the partition function for the minimum entropy problem (see [6]).

tMinimizer A wrapper to the L-BFGS-B code that has been translated from Fortran to C (see [82]).

There are other extensions, not listed here, that deal with the generation of curves and their output as images in the PNG format.

A The Network Application MtgClt/MtgSvr

The accompanying CD contains the network application MgtClt/MtgSvr, demonstrating the algorithms presented in this book. MtgClt is a Java program that is loaded from the command line shell or as an applet from an HTML page. MtgSvr is a server program based on the C++ library MtgLib to which MtgClt connects.

MtgClt and MtgSvr can be run on the same computer. On Windows, clicking on or starting the file

    \Mtg\Win32\go.bat

on the accompanying CD starts both server and client. Manually, this can be accomplished by opening a command shell and typing

    D:\Mtg\Win32>start MtgSvr\Release\MtgSvr -log con
    D:\Mtg\Win32>cd MtgClt
    D:\Mtg\Win32\MtgClt>java MtgClt

MtgClt contains a GUI (graphical user interface) that lets the user enter data in three categories:

- In the *portfolio* category, up to eight vanilla, barrier, American and customized options can be entered. Preconfigured option types include options with linear and digital payoff.
- In the *scenario* category, model coefficients such as volatility, interest rate and dividend rate (respectively foreign interest rate) are specified. All coefficients can have term structure format (tStepDrift and tStepVol are used to represent the coefficients). In addition, the volatility may exhibit uncertainty.

  The user also selects between the worst-case volatility scenario and the volatility shock scenario. In the latter case duration, periodicity and frequency are entered. Another data field determines whether the global point of view is sell-side or buy-side oriented. The distinction has been briefly made in Sect. 11.5.
- In the *advanced settings* category, finite difference scheme (explicit or Crank-Nicholson), trimming parameters (see Sect. 11.6.1), speedup techniques for American options (maintaining/collapsing of corridors of uncertainty) and time steps are selected.

A The Network Application MtgClt/MtgSvr

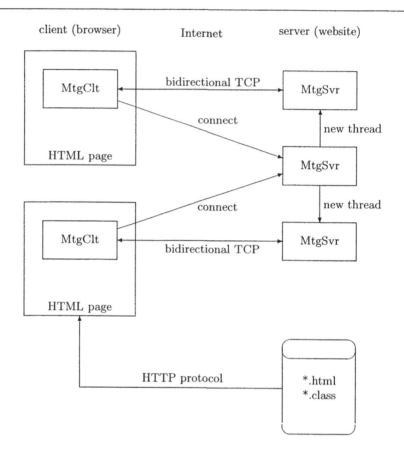

Fig. A.1. The architecture of MtgClt/MtgSvr. MtgSvr accepts incoming connections and processes requests in separate threads (Windows NT) or forked-off processes (Unix). The *.html and *.class files are served by a web server, or loaded from a command line shell

In order to give a better idea of the convergence behavior of the program for the particular pricing problem, more than one time step can be entered. The result is then computed and listed for all time steps.

Graphically, MtgClt distinguishes between "One-Click" mode in which all categories (slightly down-sized) are combined in a single entry form, and "Wizard" mode in which each category is assigned its own form. Only the Wizard mode support volatility shock scenarios and drift/volatility term structures.

Once all entries have been made, the user presses the "Start" button and MtgClt connects to MtgSvr via TCP. Because MtgClt establishes its TCP

A The Network Application MtgClt/MtgSvr   199

connection to MtgSvr on a non-standard port, both client and server must be on the same side of any firewall. By default, MtgClt assumes the server is located on the local host, "127.0.0.1".

MtgSvr handles the incoming connection by creating a new thread (under Windows NT) or by forking off a copy of itself (under Unix). Thus, several incoming requests can be handled at the same time. This architecture is shown in Fig. A.1.

MtgClt converts data entered in the entry fields into MtgScript (an example is shown in Fig. 11.1; see also Appendix B) and transmits the script to MtgSvr. MtgSvr parses the script, creates the objects it defines and executes all `evaluate` statements (of which there must be at least one). The result is sent back to MtgClt as soon as it becomes available. MtgClt/MtgSvr therefore use a simply request/response scheme to communicate.

The *Examples* submenu of MtgClt contains items that demonstrate the format in which data should be entered. The submenu is sensitive to the data category that is currently active; there are different examples for the portfolio, scenario, and advanced settings data forms.

In particular,

– All dates are entered as number of days, i.e. "30" or "60."
– All rates and volatilities are in percentages, i.e. "3.5" or "15."
– For volatilities, uncertainty is either specified by entering a range in the format
    10-20
  directly, or by entering a volatility band in the "Band (+/-)" field that is applied to all singleton data samples:
    5
  specifies a symmetric band of ±5 %, whereas
    +5/-10
  specifies an asymmetric band.
– For volatility shock scenarios, the volatility band and prior volatility are specified as triple min-prior-max volatility:
    10-12-20
  If the prior volatility is missing the average volatility is taken to be the prior. If a volatility band is entered, the prior volatility is the singleton data point, and the min and max volatility are determined by the band.
– Term structures of rates and volatilities are entered by specifying data samples column-by-column in the scenario category. Term structures are always piecewise linear. The special sample date "*" is automatically replaced with the last maturity in the options portfolio.

Figures A.2, A.3 and A.4 contain some screen snapshots of MtgClt in action.

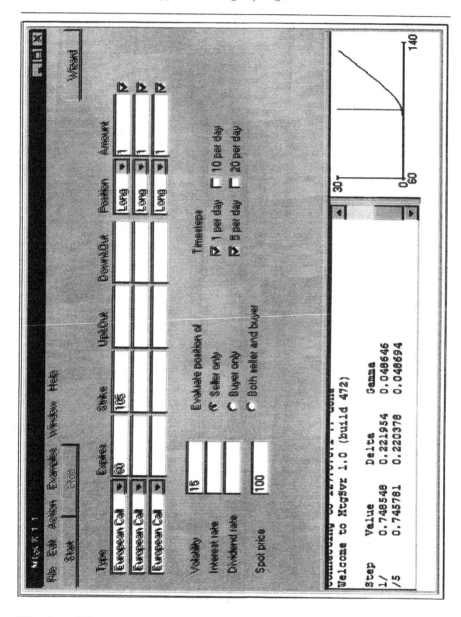

Fig. A.2. A European call option, evaluated with $dt = 1/365$ and $dt = 1/(5 \times 365)$

A The Network Application MtgClt/MtgSvr

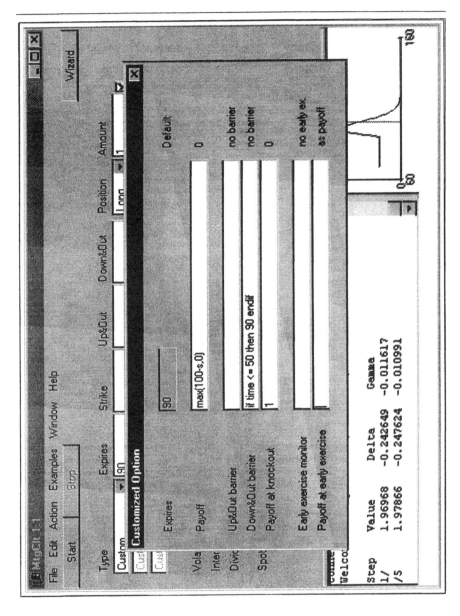

Fig. A.3. A customized 90-day down-and-out put. The barrier exists only for 50 days. The knock-out premium is $1

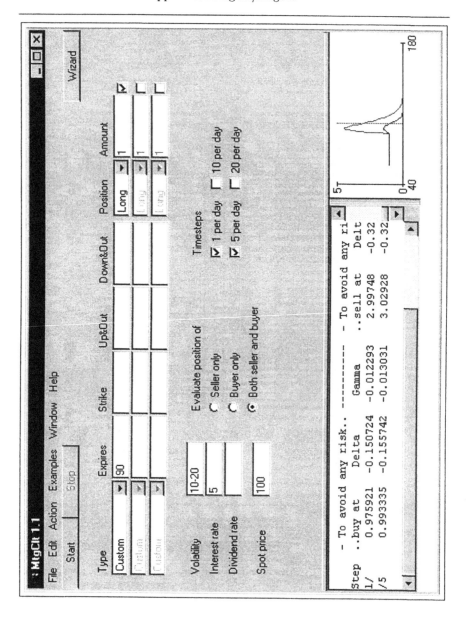

Fig. A.4. The put of Fig. A.3 evaluated under the worst case scenario in which $0.1 \leq \sigma \leq 0.2$.

B The Scripting Language MtgScript

This chapter describes the syntax of Mtg's own scripting language, MtgScript. MtgScript is used to describe external objects in the library MtgLib. (The existence of MtgScript is a bit unfortunate: a new implementation would rely on a format based on XML.)

MtgScript is ubiquitous: the component MtgSvr reads its input from a file or socket connection in this scripting language; the component MtgClt converts input made in a Java GUI into this scripting language; and the component MtgMath converts Mathematica expressions into this language (see Chapter C).

The program MtgSvr, whose server function is described in Appendix A, parses and evaluates MtgScript files if used as follows:

```
D:\Mtg\Win32>cd examples
D:\...\examples>..\MtgSvr\Release\MtgSvr MtgScript01.txt
```

It prints total value, delta, gamma, and running time for the objects in the file.

There is generally a one-to-one mapping between objects in MtgScript and the C++ classes described in Chapter 11. Classes whose instances have counterparts in MtgScript usually have a member function of the prototype

```
1  static tRetCode parse( tParser& Parser,
2      tSystem& System, tObject*& pObj );
```

This function creates a new object of the containing class and returns it in pObj.

In MtgScript, objects are generally defined by category and name:

```
1  claim a1 {
2      // properties of a1
3  }
```

The category is one of factor, claim, portfolio, vol, drift, model, scenario, lattice, path_space, bootstrap, curve, image, or optimizer. In the example, the category is claim, and its name is a1.

B The Scripting Language MtgScript

If the category is too general the object may have a type:

```
1 scenario w1 {
2    type worst_case,
3    // other properties of w1
4 }
```

Possible types for scenarios, for example, are `worst_case` or `shock`.

After an object has been defined, it can be referenced:

```
1 portfolio p1 {
2    a1 long 5
3 }
```

In this case, the portfolio called p1 consists of a long position of 5 contracts in claim a1.

Evaluators form a special object category that does not require a name. Evaluaters perform actual work: when defined they trigger a pricing operation or optimization and print out the answer:

```
evaluate {
   model m1,
   lattice l1,
   scenario w1,
   portfolio p1
}
```

The method used for the pricing operation or calibration depends on the objects that are referenced in the evaluator.

In the following sections the symbol "|" indicates an alternative value for a type or property.

B.1 Factor Objects

Factor objects are place holders. In the current implementation of MtgScript they are propertyless and only have a name. Their purpose is to make the interpretation of factors in a multi-dimensional model unambiguous.

At this point, however, Mtg only supports one-dimensional lattice models. For all lattice models the following definition can be included mechanically in every script:

```
factor s {}
```

The letter s is used arbitrarily to indicate the stock price process $S = \{S_t\}$.

Vasicek and HJM models may have multiple factors:

```
factor x1 {}
factor x2 {}
factor x3 {}
```

B.2 Claim and Portfolio Objects

Claims comprise options and US Treasury bonds. The type of a claim must be specified before its other properties.

```
1  claim a1 {
2      type european_call
3         | european_put
4         | american_call
5         | american_put
6         | ust_bond
7         | custom,
8      // type-specific properties of a1
9  }
```

We will see that the types and type-specific properties of the instrument directly correspond to the implementation aspects described in Sect. 11.1.

Options of type european_call, european_put, american_call and american_put are defined as follows:

```
1  claim a2 {
2      type european_call | ...,
3      maturity 60,
4      strike 100,
5      down_and_out 90,
6      up_and_out 110,
7      payoff linear | digital
8  }
```

The maturity T is in days. The barriers down_and_out and up_and_out are only required for barrier options and can appear in any combination; the barriers are always flat. If no payoff value is given, the payoff is automatically linear, corresponding to standard vanilla calls and puts. A digital payoff yields 1 if $S_T \geq$ strike, and 0 otherwise.

Instruments of type custom allow to specify individual payoff and monitoring expressions by defining the following properties:

```
1  claim a3 {
2      type custom,
3      maturity 60,
4      payoff <numerical expr>,
5      monitor <exercise expr>,
6      exercise_payoff <numerical expr>,
7      up_and_out <numerical expr>,
8      down_and_out <numerical expr>,
9      knockout_payoff <numerical expr>
10 }
```

Each of these properties is optional. A numerical expression can use any of the quantities time, day, fraction_of_day, or any factor defined as factor object. Conditional expressions are written

        if ... then ... else ... endif

The min and max functions are also available. Examples of custom claims are:

```
 1  factor s {}
 2
 3  claim a4 {
 4      type custom,
 5      maturity 60,
 6      payoff max( s - 100, 0 ),
 7      up_and_out 110,
 8      down_and_out if time >= 30 then 90 else 80 endif,
 9      knockout_payoff 0
10  }
```

This example is a double-barrier call option with strike 100. The down-and-out barrier is time dependent; after 30 days it drops by 10 points. The knockout_payoff is paid when the barrier is reached; in this case, nothing is paid.

The monitor expression controls early exercise as defined in Sect. 11.1. It may return one of the values force_exercise, may_exercise or dont_exercise, corresponding to the exercise region, region of uncertainty, and continuation region of Sect. 8.2. A regular American option would be defined by indicating that the scenario must select the exercise policy when it is optimal:

```
1  factor s {}
2
3  claim a5 {
4      type custom,
5      maturity 60,
6      payoff max( s - 100, 0 ),
7      monitor may_exercise,
8      exercise_payoff max( s - 100, 0 )
9  }
```

It is possible to define exotic barrier options through the monitor as well:

```
1  factor s {}
2
3  claim a6 {
4      type custom,
5      maturity 60,
6      payoff max( s - 100, 0 ),
```

B.2 Claim and Portfolio Objects   207

```
7    monitor if s > 100 + time then
8        force_exercise else dont_exercise,
9    exercise_payoff 0
10 }
```

Only barriers defined by the properties up_and_out and down_and_out are used to align the levels of the lattice in the algorithm of Fig. 5.4.

Regular and custom options support all the algorithms in Chapters 6, 7 and 8. In MtgLib they are priced on a lattice. The instrument type ust_bond, on the other hand, is used for Monte-Carlo calibration of the discount curve based on the theory of Sect. 4.3. The name ust_bond is short for US Treasury bond; here is the definition:

```
1  claim a7 {
2      type ust_bond,
3      maturity "6 Nov 1994",
4      principal 100,
5      coupon 5,
6      settlement following 2
7               | preceding 2
8               | modified
9               | fixed "8 Nov 1994"
10              | identity,
11     bid_price 101,
12     ask_price 102
13 }
```

Principalprincipal and coupon are both in absolute amounts. Calendar dates can be entered in one of the following formats:

```
11/6/1994
Nov 6 1994      Nov 6, 1994      Nov/6/1994
6 Nov 1994      6-Nov-1994       6/Nov/1994
```

The bad-businessday conventions following and preceding are followed by the number of businessdays. If that number is non-zero the current system date is moved by the specified number of businessdays in the desired direction. If the number is missing or zero the system date is only adjusted, not moved. modified stands for modified-following. In MtgLib, only weekends are bad businessdays.

Instead of a bid-ask spread via bid_price and ask_price a single price can be specified, through the alternative property price.

The special instrument type custom also supports fixed-income instruments, though only for Monte-Carlo calibration. The following example shows how fixed cashflows are defined in a custom instrument.

```
1  claim a8 {
2    type custom,
3    maturity "6 Nov 1994",
4    cashflow "6 Nov 1993" <numerical expr>,
5    cashflow "6 Nov 1994" <numerical expr>,
6    ...
7    bid_price 101,
8    ask_price 102
9  }
```

The numerical expression may use `time`, `day`, `fraction_of_day`, or any `factor`. Conditional expressions and `min` and `max` are also allowed.

Instruments are combined into portfolios as follows:

```
1  portfolio p1 {
2    a1 long 1,
3    a2 short 2,
4    ...
5  }
```

If the position quantity is missing, one is assumed. The position modifiers `long` and `short` can be replaced with the neutral modifier `position`, in which case the quantity must be positive for long positions, and negative for short positions.

B.3 Model, Drift and Volatility Objects

Model objects describe stochastic models. Mtg supports three models: Black-Scholes for lattice pricing, short-rate Vasicek and multi-factor HJM for Monte-Carlo pricing and calibration.

Models are defined as follows:

```
1  model m1 {
2    type bs
3         | vasicek
4         | hjm_gaussian,
5    days_per_year 365,
6    // type-specific properties of m1
7  }
```

The property `days_per_year` is defined for every model and controls the time scale based on which instantaneous rates and volatilities are compounded over time periods. A year in which only business days count, for example, can be approximated by setting this parameter to 255.

B.3 Model, Drift and Volatility Objects

The Black-Scholes model is the only type of model that can be used for lattice pricing of portfolios of vanilla, barrier and American options. It has these properties:

```
1  model m2 {
2      type bs,
3      days_per_year 365,
4      discount <object name>,
5      carry <object name>,
6      vol <object name>,
7      factor <object name> 100
8  }
```

The `discount` drift object models the term structure of the interest rate r_t. The `carry` drift object models the term structure of the dividend rate q_t (equity options) or the foreign interest rate (FX options). The volatility object referenced by `vol` models the term structure of volatility, including bands of uncertainty. The factor is usually called `s`; the name is followed by the initial value.

For FX options, the keywords `discount` and `carry` can be replaced with `domestic` and `foreign`, respectively.

Vasicek and HJM models are used the calibration, as described in Sect. 4.3.

The Vasicek short-rate model (defined in Sect. 3.1.3) has the following properties:

```
1  factor r1 {}
2
3  model m3 {
4      type vasicek,
5      days_per_year 365,
6      factor r1 sigma 0.05 theta 1.0 alpha 0.2,
7      mean <object name>
8          | 0.05,
9      initial <object name>
10         | 0.05
11 }
```

The `mean` property either references a time-dependent drift or bootstrapping object, or is set to a constant number. The factor coefficients `sigma`, `theta` and `alpha` are constant and corresponds to the coefficients of (3.39) as follows:

$$\sigma_t = \text{sigma}$$
$$\theta_t = \text{theta} \times \text{mean}_t \qquad \text{(B.1)}$$
$$\alpha_t = \text{alpha}$$

Here, mean_t is used to denote the mean level at time t. The property time-dependent nature of `mean` makes the level of mean reversion time-dependent.

The initial forward rate is specified in the property `initial`, which can be a constant, a drift term structure, or a bootstrapping object. If `initial` is not constant the object's value at time $t = 0$ (the base date) is extracted (the rest of the object is ignored).

The HJM model is Gaussian and has the following properties:

```
1  factor x1 {}
2  factor x2 {}
3
4  model m4 {
5      type hjm_gaussian,
6      days_per_year 365,
7      factor x1 sigma 0.05 kappa 0.2,
8      factor x2 sigma 0.15 kappa 0.5,
9      ...,
10     initial <object name>
11            | 0.05
12 }
```

The number of factors is not limited. Factors are independent and follow the Gaussian process

$$\mathrm{d}X^i = \mathtt{kappa}_i \, \mathrm{d}t + \mathtt{sigma}_i \, \mathrm{d}Z^i \qquad (B.2)$$

The initial forward rate curve is specified in the property `initial`, which can be a drift term structure, a bootstrapping object, or a constant.

Drift objects are defined as a term structure of `implied` or `forward` rates. Their instantaneous rates are always constant between samples. For example,

```
1  drift r1 {
2      implied 30 0.05,
3      implied 60 0.06,
4      implied 90 0.07
5  }
```

and

```
1  drift r2 {
2      forward 30 0.05,
3      forward 60 0.07,
4      forward 90 0.09
5  }
```

both define the following term structure of the short rate:

$$r_t = \begin{cases} 0.05 & \text{if } 0 \leq t \leq 30 \text{ days} \\ 0.07 & \text{if } 30 < t \leq 60 \\ 0.09 & \text{if } 60 < t \leq 90 \end{cases} \qquad (B.3)$$

B.3 Model, Drift and Volatility Objects

The relation between `implied` or `forward` rates is as follows:

$$\text{implied}_t = \frac{1}{t} \int_0^t \text{forward}_u \, du \tag{B.4}$$

The drift object can contain both `implied` and `forward` specifications at the same time.

Volatility objects are defined as a term structure of `implied` or `forward` volatilities. The spot volatility is always constant between samples. Volatility objects are defined as follows:

```
1  vol v1 {
2      implied 30 0.1,
3      implied 60 0.2,
4      forward 90 0.2
5  }
```

The relation between `implied` or `forward` volatility specifications is as follows:

$$\text{implied}_t = \sqrt{\frac{1}{t} \int_0^t \text{forward}_u^2 \, du} \tag{B.5}$$

The volatility object can contain both `implied` and `forward` specifications (for different dates!) at the same time.

Uncertain volatility scenarios require that volatility objects specify uncertainty. Uncertainty is expressed as a range or as a range with a prior volatility (for volatility shock scenarios). Without an explicit prior the minimum must come before the maximum, and the prior is automatically set to the average:

```
1  vol v2 {
2      implied 30 0.1 .. 0.2,
3      implied 60 0.2 .. 0.3,
4      forward 90 0.2 .. 0.3
5  }
```

With a prior the order is minimum before prior before maximum:

```
1  vol v3 {
2      implied 30 0.1 .. 0.15 .. 0.2,
3      implied 60 0.2 .. 0.2  .. 0.3,
4      forward 90 0.2 .. 0.3  .. 0.3
5  }
```

The minimum, prior and maximum volatilities are converted separately into three independent term structures.

Both drift and volatility objects are based on the same underlying implementation of piecewise constant term structure objects. See Sect. 11.4 for more information.

B.4 Lattice and Path Space Objects

While model objects define the mathematical properties of stochastic processes, lattice and path space objects define the concrete properties of the numerical implementation.

Lattice objects are used for the pricing of portfolios of vanilla, barrier and American options under a Black-Scholes model. They are built with the algorithms of Chapter 5 and defined as follows:

```
1 lattice l1 {
2     shape tree 3.5
3         | box 3.5,
4     method implicit
5         | explicit,
6     model m1,
7     portfolio p1,
8     time_step 1
9 }
```

If the shape property is set to tree, the lattice is a trinomial tree with three root nodes. If the shape property is set to box, the size of the lattice does not change over time. In both cases the number after the shape property indicates the number of standard deviations after which the lattice is trimmed (there is no trimming if this number is missing). The value 3.5 is a good default.

The property method specifies the backwards induction scheme. The implicit method refers to the Crank-Nicholson finite-difference scheme. The shape/method combinations tree/explicit respectively box/implicit make most sense.

The lattice object must be tied to a Black-Scholes model. It must be tied to a portfolio because the latest maturity date in the portfolio determines the number of time steps. The size of the time_step is specified in days.

In order to make a model usable for Monte-Carlo pricing or calibration it must be completed with a path space. A path space is defined as follows:

```
 1 path_space t1 {
 2     base "1/1/2001",
 3     day_count dc_30e_360
 4         | dc_30_360_isda
 5         | dc_30_360_psa
 6         | dc_30_360_sia
 7         | dc_act_360
 8         | dc_act_365
 9         | dc_act_365_25
10         | dc_act_365_isda
11         | dc_act_365_nl
```

```
12            | dc_act_act,
13      scale day
14            | year 4,
15      model m2,
16      portfolio p2,
17      size 1000,
18      seed 1234567890 123456789,
19      // weight property
20      // save property
21 }
```

Events on individual paths are associated with actual dates. The path space must have a base date and a day_count convention (the choices are according to [74]). The scale property corresponds to the time_step property for lattices. It determines the time interval in which random samples are taken: day means daily, year means a specified number of times per year.

The path space must link to a model and a portfolio. The size of the path space is the number of Monte-Carlo paths. The seed property is optional. It seeds the random number generator.

The weight property is optional. If present it holds the result of a calibration run for the path space (recall that calibration assigns new weights to Monte-Carlo paths). Weights are either stored in a file or listed in a block, i.e. in this format

    weight "filename"

or this format

```
1 weight {
2     0.1, 0.1, 0.1, 0.1, 0.1,
3     0.1, 0.1, 0.1, 0.1, 0.1
4 }
```

If the weight property is missing each Monte-Carlo path has equal weight.

The save property is also optional. If present the path space is saved to the specified file after calibration. It can then be used for pricing. The save property is defined as follows:

    save "filename"

B.5 Bootstrapping, Curve and Image Objects

Bootstrapping and curve objects are auxiliary objects that assist in the calibration of Vasicek and HJM models. Image objects display bootstrapping and curve objects visually.

Bootstrapping objects take a portfolio of fixed-income instruments with different maturities and convert them into a piecewise constant short rate curve that re-prices those instruments. The instruments must have the `price` property.

Bootstrapping objects are defined as follows:

```
1  bootstrap b1 {
2     portfolio p1,
3     save "filename",
4     save leg "filename",
5     save gnu_data "filename",
6     save gnu_tics "filename"
7  }
```

The `save` property is optional, in all its variations. If present the resulting curve is saved in these formats:

- The original format (no format specifier) outputs the entire bootstrap object, in MtgScript. It can then be used in another script as a bootstrapping or curve object.
- The `leg` format consists of a list of maturity dates and yields.
- The `gnu_data` format consists of a specification of the curve that can be plotted by the program Gnuplot.
- The `gnu_tics` format consists of a specification of axis tics for the program Gnuplot. It is concerned with the formatting of maturity dates.

Curve objects capture the results of calibration under Monte-Carlo. They are connected to a path space and instantiated after the weights of the path space have been calibrated. Curves can also be defined explicitly and used as input for the HJM model.

Curves that are used to capture calibration results are defined as follows:

```
1  curve c1 {
2     type forward
3        | yield,
4     path_space t1,
5     base "1/1/2001",
6     day_count ...,
7     scale day
8        | year 4,
9     save "filename",
10    save sample "filename",
11    save drift "filename",
12    save gnu_data "filename",
13    save gnu_tics "filename"
14 }
```

Two types of curves are supported: `forward` curves capture the short rate r_t as a function of time; `yield` curves describe the yield of a zero-coupon bond as a function of its maturity. In both cases the calibrated path space is used to compute the curve samples. For example, the yield obtained from the curve and the value of a zero-coupon bond, priced under the calibrated path space, correspond to each other (assuming continuous compounding).

The properties `base`, `day_count` and `scale` are optional. If missing the respective properties of the `path_space` are inherited. (By choosing a `scale` that is larger than the `scale` of the path space the calibrated curve is effectively smoothed.)

The `save` property is optional, in all its variations. If present the curve is saved in these formats:

– If no format specifier is given the entire curve object is written in MtgScript. The reference to the path space is replaced with actual sample values.
– The `sample` format consists of a list of sample rates or yields only.
– The `drift` format stores the curve as a `drift` object.
– The `gnu_data` format consists of a specification of the curve that can be plotted by the program Gnuplot.
– The `gnu_tics` format consists of a specification of axis tics for the program Gnuplot.

While curve objects can be save as and thus converted into drift objects, the opposite way is not supported.

Curves can also be defined explicitely as follows:

```
1  curve c2 {
2      type forward
3          | yield,
4      base "1/1/2001",
5      day_count ...,
6      scale day
7          | year 4,
8      sample {
9          0.05, 0.055, 0.06,
10         0.065, 0.07, 0.07
11     }
12 }
```

The spacing of the sample values is determined from the `base` date and the `scale`. This is the format in which the `save` property without format specifier writes the curve.

Image objects combine one or more bootstrapping or curve objects into an image in the PNG format. They are defined as follows:

```
1  image g1 {
2      title "My image",
3      portfolio p1,
```

```
 4    size <x-size> <y-size>,
 5    min <min sample value>,
 6    max <max sample value>,
 7    curve c1 red lines,
 8    curve c2 green lines,
 9    bootstrap b1 blue steps,
10    ...,
11    save "filename"
12 }
```

The `portfolio` property is optional. If present individual instrument maturities are indicated by vertical bars in the graph. The `size` property is in pixels. If `min` and/or `max` are missing the sample range is taken from the data.

The number of curves or bootstrapping objects in a single image is not limited. Only three colors are avaliable: `red`, `green` and `blue`. Curves should be drawn in the `lines` style, due to their continuous nature. Bootstrapping object should be drawn in the `steps` style.

The image is created and saved to the file specified in the `save` property.

B.6 Scenario and Optimizer Objects

Scenario objects are needed for lattice pricing. Optimizer objects are used for Monte-Carlo pricing and calibration, in which case the minimum-entropy approach already implies a scenario.

Scenario objects for lattices are defined as follows:

```
1 scenario w1 {
2    type worst_case
3         | shock,
4    position buyer
5         | seller,
6    // type-specific properties of w1
7 }
```

Worst-case scenarios have been introduced in Sect. 4.2.1. Volatility shock scenarios have been discussed in Chapter 9. Both types of scenarios distinguish between a `seller`'s and a `buyer`'s position.

Worst-case scenarios do not have type-specific properties. Volatility shock scenarios have the following type-specific properties:

```
1 scenario w2 {
2    type shock,
3    position buyer
4         | seller,
5    duration 10,
```

```
6      periodicity 1,
7      repetitions 1
8 }
```

The parameters duration, periodicity and repetitions correspond to the quantities d, p and f of Def. 22. The duration and periodicity properties are specified in days.

Optimizer objects are used to calibrate models. Currently Mtg only supports minimum-entropy optimization for the Vasicek and HJM models under Monte-Carlo. The optimizer is defined as follows:

```
1 optimizer o1 {
2      type entropy,
3      low -10,
4      high 10
5 }
```

The optimizer uses the method of Lagrange multipliers outlined in Sect. 4.3. The low and high bounds define a bounding box for the Lagrange multipliers $\lambda_1, \ldots, \lambda_{\bar{k}}$. If no solution within the bounding box is found the optimization (and therefore the calibration of the model) fails.

B.7 Evaluation Objects and Examples

Mtg supports two kinds of evaluation:

1. Evaluation of a portfolio of vanilla, barrier and American equity or FX options on a lattice produces a price.
2. Evaluation of a portfolio of fixed-income instruments (ust_bond and some instances of custom claims) under Monte-Carlo produces a price if no optimizer is present, and a calibrated path space if an optimizer is present.

   Evaluation on a lattice is straightforward:evaluate

```
1 // all object definitions
2
3 evaluate {
4      portfolio p1,
5      model m1,
6      lattice l1,
7      scenario w1,
8      accuracy exact | low
9 }
```

218   B The Scripting Language MtgScript

All objects must have been defined. The evaluation object does not need a name; it immediately produces a result. The property accuracy is only relevant for portfolios that contain American options. It selects between maintaining the corridor of uncertainty (exact; see Sect. 8.2.1) and collapsing the corridor of uncertainty (low; see Sect. 8.2.2).

Evaluation under Monte-Carlo is equally straightforward:

```
1  system {
2      base "1/1/2001"
3  }
4
5  // all object definitions
6
7  evaluate {
8      portfolio p1,
9      model m2,
10     path_space t1,
11     optimizer o1,
12     image g1,
13     image g2,
14     ...
15 }
```

Optimizers and images are optional. If no optimizer is present no calibration is performed. Only images listed in the evaluation object are generated. All objects must have been defined before their use in an evaluation object.

The unnamed system object should be the first object in the script. It defines the default base date; objects that do not specify a base date will take it from this object. (A system date is not needed for lattice evaluation because only Monte-Carlo pricing and calibration uses calendar dates.)

In the first example we evaluate a portfolio of 4 option positions on a lattice, first under a worst_case scenario with $\sigma_{min} = 0.1$ and $\sigma_{max} = 0.2$, then under a volatility shock scenario with prior $\sigma_0 = 0.15$.

```
1  // a1/a2 form a spread:
2
3  claim a1 {
4      type european_call,
5      maturity 60,
6      strike 100,
7      payoff linear
8  }
9
10 claim a2 {
```

```
11      type european_call,
12      maturity 60,
13      strike 110,
14      payoff linear
15 }
16
17 // A knock-out put:
18
19 claim a3 {
20      type european_put,
21      maturity 60,
22      strike 95,
23      down_and_out 85,
24      payoff linear
25 }
26
27 // An American put:
28
29 claim a4 {
30      type american_put,
31      maturity 50,
32      strike 100,
33      payoff linear
34 }
35
36 portfolio p1 {
37     a1 long,
38     a2 short,
39     a3 long,
40     a4 short
41 }
42
43 // (Domestic) Interest rate:
44
45 drift r {
46      implied 60 0.055
47 }
48
49 // Dividend/foreign interest rate:
50
51 drift q {
52      implied 60 0.025
53 }
54
```

```
55  // Minimum vol is 10%.
56  // Prior vol is 15%.
57  // Maximum vol is 20%.
58
59  vol v {
60      implied 60 0.1 .. 0.15 .. 0.2
61  }
62
63  factor s {
64  }
65
66  model m1 {
67      type bs,
68      days_per_year 365,
69      discount r,
70      carry q,
71      vol v,
72      factor s 100
73  }
74
75  // Explicit finite differencing, trimming
76  // after 3.5 standard deviations, 10 time
77  // steps per day:
78
79  lattice l1 {
80      shape tree 3.5,
81      method explicit,
82      model m1,
83      portfolio p1,
84      time_step 0.1
85  }
86
87  scenario w1 {
88      type worst_case,
89      seller
90  }
91
92  evaluate {
93      portfolio p1,
94      model m1,
95      lattice l1,
96      scenario w1,
97      accuracy exact
98  }
```

```
 99
100  // Result: 1.63631
101
102  // Now evaluate the same portfolio under a
103  // shock scenario - prior vol is 15%.
104
105  scenario w2 {
106      type shock,
107      seller,
108      duration 3,
109      periodicity 1,
110      repetitions 1
111  }
112
113  evaluate {
114      portfolio p1,
115      model m1,
116      lattice l1,
117      scenario w2,
118      accuracy exact
119  }
120
121  // Result: 1.03289
```

In the next example, four US Treasury bonds are used to calibrate a Vasicek model under Monte-Carlo. The calibration result is captured in the image calibration02.png, together with the piecewise-constant forward rates obtained from the bootstrapping object. The calibrated path space is stored in pathspace02.txt.

```
 1  system {
 2      base "15 Jan 2000"
 3  }
 4
 5  // Four US Treasury bonds. The
 6  // coupons are absolute amounts.
 7
 8  claim a1 {
 9      type ust_bond,
10      maturity "15 Jan 2005",
11      settlement identity,
12      principal 100,
13      coupon 4.5,
14      price 98
15  }
16
```

```
17  claim a2 {
18      type ust_bond,
19      maturity "15 Jan 2007",
20      settlement identity,
21      principal 100,
22      coupon 4.5,
23      price 100
24  }
25
26  claim a3 {
27      type ust_bond,
28      maturity "15 Jan 2010",
29      settlement identity,
30      principal 100,
31      coupon 5,
32      price 100
33  }
34
35  claim a4 {
36      type ust_bond,
37      maturity "15 Jan 2030",
38      settlement identity,
39      principal 100,
40      coupon 5.5,
41      price 105
42  }
43
44  // Although we need to define a portfolio,
45  // the quantities won't matter for bootstrapping
46  // or calibration.
47
48  portfolio p1 {
49      a1 long,
50      a2 long,
51      a3 long,
52      a4 long
53  }
54
55  // Compute a piecewise constant forward
56  // rate curve via simple bootstrapping.
57
58  bootstrap b1 {
59      portfolio p1,
60      save "bootstrap02.txt"
```

```
61  }
62
63  factor r { }
64
65  // Use the bootstrapped step function as
66  // mean for the Vasicek model. The short
67  // rate at the base date is 5% ("initial").
68
69  model m1 {
70      type vasicek,
71      days_per_year 365,
72      factor r sigma 0.01 theta 1.0 alpha 1.0,
73      initial 0.05,
74      mean b1
75  }
76
77  // The base date is taken from the
78  // system object. Time steps are
79  // spaced monthly.
80
81  path_space t1 {
82      day_count dc_act_365,
83      scale year 12,
84      model m1,
85      portfolio p1,
86      size 50000,
87      save "pathspace02.txt"
88  }
89
90  optimizer o1 {
91      type entropy,
92      low -100,
93      high 100
94  }
95
96  // Save the resulting forward rates and yields
97  // so that other programs can read them for
98  // simple bond math.
99
100 curve c1 {
101     type forward,
102     path_space t1,
103     save "forward02.txt"
104 }
```

```
105
106 curve c2 {
107     type yield,
108     path_space t1,
109     save "yield02.txt"
110 }
111
112 // Put both curves and the bootstrapping
113 // result into an image.
114
115 image g1 {
116     title "Calibration result",
117     portfolio p1,
118     size 1024 512,
119     curve c1 red lines,
120     curve c2 green lines,
121     bootstrap b1 blue steps,
122     save "calibration02.png"
123 }
124
125 // Calibrate the pathspace t1 and save
126 // the result in "pathspace02.txt".
127
128 evaluate {
129     portfolio p1,
130     model m1,
131     path_space t1,
132     optimizer o1,
133     image g1
134 }
```

In the third and last example the benchmark bonds of example 2 are repriced under the calibrated path space. As expected, the value of portfolio p1 is 403. Of course, any other instrument could have been priced on the calibrated path space as well.

```
 1 system {
 2     base "15 Jan 2000"
 3 }
 4
 5 // The same four US Treasury bonds that
 6 // were used to calibrate the Vasicek
 7 // model:
 8
 9 claim a1 {
10     type ust_bond,
```

```
11     maturity "15 Jan 2005",
12     settlement identity,
13     principal 100,
14     coupon 4.5,
15     price 98
16 }
17
18 claim a2 {
19     type ust_bond,
20     maturity "15 Jan 2007",
21     settlement identity,
22     principal 100,
23     coupon 4.5,
24     price 100
25 }
26
27 claim a3 {
28     type ust_bond,
29     maturity "15 Jan 2010",
30     settlement identity,
31     principal 100,
32     coupon 5,
33     price 100
34 }
35
36 claim a4 {
37     type ust_bond,
38     maturity "15 Jan 2030",
39     settlement identity,
40     principal 100,
41     coupon 5.5,
42     price 105
43 }
44
45 portfolio p1 {
46     a1 long,
47     a2 long,
48     a3 long,
49     a4 long
50 }
51
52 // Include the calibration results of example 2.
53
54 include "bootstrap02.txt"
```

```
55
56 // Re-create the model. Note that b1 is
57 // defined in bootstrap02.txt
58
59 factor r { }
60
61 model m1 {
62    type vasicek,
63    days_per_year 365,
64    factor r sigma 0.01 theta 1.0 alpha 1.0,
65    initial 0.05,
66    mean b1
67 }
68
69 // Include the calibration results of example 2.
70
71 include "pathspace02.txt"
72
73 // Re-price the benchmark bonds. Note that
74 // path space t1 is defined in pathspace02.txt.
75
76 evaluate {
77    portfolio p1,
78    model m1,
79    path_space t1
80 }
```

C Mathematica Extensions

The component MtgMath consists of two parts:

- The external program MtgMath.exe contains MtgLib and Mathematica's MathLink interface package.
- The add-on package MtgMath.m contains definitions of Mathematica functions that translate between Mathematica expressions and Mtg's scripting language.

MtgMath is loaded in Mathematica with

    Needs["RB'MtgMath'"]

After MtgMath has been loaded it is used as follows:

- Symbolic Mathematica expressions represent Mtg objects for volatility, drift, portfolios, and scenarios.
- A collection of Mtg objects is converted into an Mtg script (whose syntax is described in Chapter B).
- This script is run to obtain a price.

This chapter defines the syntax of object expressions and the synopsis of the conversion and pricing functions. See Fig. 1.2 for a concrete example.

C.1 The Syntax of Object Expressions

In the following, all Mathematica expressions are shown literally, with the exception of the symbol "|" that is used to denote alternative flags.

Volatility objects are defined as follows:

```
1 VOL[IMPLIED[maturity, annual rate], ...]
2
3 VOL[IMPLIED[maturity,
4     {min rate, max rate}], ...]
5
6 VOL[IMPLIED[maturity,
7     {min rate, prior rate, max rate}], ...]
```

The maturity is in days. The annual volatility is a percentage. A sequence of IMPLIED terms with increasing maturities creates a volatility term structure with piecewise constant instantaneous forward volatilities. The prior rate in the third version is needed for volatility shock scenarios.

Drift objects are defined as follows:

```
DRIFT[IMPLIED[maturity, annual rate], ...]
```

The maturity is in days. The annual volatility is a percentage. A sequence of IMPLIED terms with increasing maturities creates a drift term structure with piecewise constant instantaneous forward rates.

Options are defined as follows:

```
1  ECALL[maturity, strike]
2
3  ECALL[maturity, strike,
4      UpAndOut->barrier,
5      DownAndOut->barrier,
6      Payoff->Linear|Digital]
7
8  ACALL[maturity, strike, ...]
9  EPUT[maturity, strike, ...]
10 APUT[maturity, strike, ...]
```

The maturity is in days. The parameters UpAndOut, DownAndOut, indicating barrier options, and Payoff, indicating linear or digital payoff, can be combined in any way; the example applies them in unison. ECALL and EPUT are European, ACALL and APUT American.

Portfolios are defined as follows:

```
1  PORTFOLIO[
2      LONG[option, quantity],
3      SHORT[option, quantity],
4      ...]
```

with an arbitrary number of option positions. If quantity is missing one is substituted.

Scenarios are defined as follows:

```
1  SCENARIO[WORSTCASE[SELLER]]
2  SCENARIO[WORSTCASE[BUYER]]
3
4  SCENARIO[SHOCK[SELLER],
5      Duration->d,
6      Periodicity->p,
7      Repetitions->f]
8
9  SCENARIO[SHOCK[BUYER],
```

```
10      Duration->d,
11      Periodicity->p,
12      Repetitions->f]
```

where the first two scenarios are of the type defined in Sect. 4.2, and the last two are shock scenarios as defined in Chapter 9. The parameters Duration, Periodicity and Repetitions correspond to the quantities d, p and f of Def. 22.

Volatility, drift, portfolio and scenario objects can be put together to form an Mtg script. The objects are collected first:

```
1  objects = {
2     VOL[...],
3     DRIFT[...],
4     DRIFT[...],
5     PORTFOLIO[...],
6     SCENARIO[...]};
```

The first drift object is always used as the risk-free rate. The second drift object, if present, is used as dividend rate. The objects are then converted into an Mtg script:

```
1  mtg = Script[objects,
2     AssetPrice->100,
3     DaysPerYear->365,
4     Shape->Tree|Box,
5     Deviations->3.5,
6     TimeStep->1,
7     Accuracy->Exact|Low,
8     Method->Explicit|Implicit,
9     Profile->FileName];
```

AssetPrice denotes S_0; the default is 100. DaysPerYear scales drift and volatility; the default is 365. Shape determines the shape of the lattice: a tree has 3 root nodes, the middle one representing S_0; a box is rectangular, with each time slice in the lattice having the same number of nodes. Deviations represents the number of standard deviations in either direction after which the lattice is trimmed; the default is 3.5. TimeStep is the number of days per time step; the default is 1. Accuracy indicates whether the speedup techniques of Chapter 8 should be applied (Low) or not (Exact). The numerical Method is either Explicit or Implicit (Crank-Nicholson).

If a file name is specified in the Profile parameter, diagnostic information is written to that file when the script is run. The graphical functions listed in Sect. C.3 depend on this information.

After a script has been created, it can be run:

RunScript[mtg]

The output has the following format:

230   C Mathematica Extensions

```
1 {
2     { total value, delta, gamma, running time },
3     { gradient w.r.t. each option },
4     { root nodes of the lattice }
5 }
```

The total value, delta and gamma refer to the entire portfolio:

$$\text{total value} = \hat{F}_0, \quad \text{delta} = \frac{\partial \hat{F}_0}{\partial S}, \quad \text{gamma} = \frac{\partial^2 \hat{F}_0}{\partial S^S} \tag{C.1}$$

Individual components are contained in the gradient as units. Assume the PORTFOLIO object contains option positions $\lambda_1, \ldots, \lambda_k$, and the output gradient is $g_1, \ldots, g_k$; then

$$\frac{\partial \hat{F}_0}{\partial \lambda_i} = g_i \quad (1 \le i \le k) \tag{C.2}$$

The number of root nodes of the lattice depends on the value of the Shape parameter. Trees have 3 root nodes; boxes as many as required for their rectangular shape. In both cases the coordinate value of S_0 as well as the value $\hat{F}_0$ is contained in the output. Nodes are always centered around the value of parameter AssetPrice.

C.2 Turning Scripts into Functions

When run with RunScript, a script only produces one price. In order to create a graph of prices, a collection of objects must be converted into a function.

Let object denote a collection of volatility, drift, portfolio and scenario objects. This collection is converted into a price function as follows:

```
1 func = MakePriceFunc[{args},
2     objects,
3     AssetPrice->100,
4     DaysPerYear->365,
5     Shape->Tree|Box,
6     Deviations->3.5,
7     TimeStep->1,
8     Accuracy->Exact|Low,
9     Method->Explicit|Implicit];
```

Here, args is a list of argument names which appear in the objects or the parameters. For example, a function that maps an asset price to a portfolio value is defined as follows:

```
1 MakePriceFunc[{s},
2    objects,
3    AssetPrice->s]
```

A function that maps a volatility to a price is defined as follows:

```
1 MakePriceFunc[{v},
2    objects without volatility,
3    VOL[IMPLIED[100, v]],
4    AssetPrice->100]
```

The resulting price function can be used like any other Mathematica function:

```
1 x = func[100];
2 Plot[func[s], {s, 90, 110}];
```

To create Mathematica functions for the delta or gamma of a collection of parameterized objects, use `MakeDeltaFunc` respectively `MakeGammaFunc`. All three functions are based on `MakeScriptFunc`.

C.3 Profiling and Diagrams

The parameter `Profile` of function `Script` sets the name of an output file into which a profile of the pricing operation is recorded during execution. This profile can be inspected and displayed graphically.

The following code reads a profile back in after pricing:

```
1 IgnoreLastSlice = False|True;
2 AlwaysGrayScale = False|True;
3 OpenFile[profile];
```

where `profile` is the file name given in `Script`'s parameter `Profile`. `IgnoreLastSlice` and `AlwaysGrayScale` are optional global flags. If the flag `IgnoreLastSlice` is true the last time slice at T_N is dropped from the profile data because it may be uninteresting to the user.

`OpenFile` prints a summary of layers and their colors. Each layer represents one lattice instance; also shown are the instruments indices and the lattice tag (see Sect. 11.5 for a definition of lattice tags).

The complexity of a particular node in a particular layer is defined as the number of early exercise combinations examined during pricing under the chosen accuracy (`Exact` or `Low`). It is graphically represented for layer n as follows:

`Show[GetComplexityPlot[n]]`

A dependency graph shows how values travel between lattice instances. Figure 7.1, for example, depicts this situation for two lattice instances and barrier options. For a given layer n, a color-coded dependency graph can be obtained with

Show[GetDependencyPlot[n]]

In this graph, each node is colored with the color of the layer from whose lattice instance it has received its values.

Definition 19 defines the corridor of uncertainty for American option portfolios. The following two expressions display regions where early exercise is uncertain and thus needs to be resolved by comparing early exercise combinations, and regions where early exercise is not combinatorial:

1 Show[GetMayExPolicyPlot[n]]
2 Show[GetForceExPolicyPlot[n]]

In both cases, n is the index of a layer.

Uncertain volatility scenarios select volatility locally, for each individual lattice node. The following expression shows the volatility selected during pricing graphically for layer n:

Show[GetVolatilityPlot[n]]

The total value $\hat{V}(j, i; n)$ of all nodes (j, i) of the lattice instance belonging to layer n can be displayed as follows

Show[GetTotal[n]]

References

1. Arnold, L. (1973): Stochastic Differential Equations: Theory and Applications. Wiley, New York
2. Avellaneda, M., Buff, R. (1998): Combinatorial Implications of Nonlinear Models: the Case of Barrier Options. Applied Mathematical Finance, **6**, 1–18
3. Avellaneda, M., Levy, A., Parás, A. (1995): Pricing and Hedging Derivative Securities in Markets with Uncertain Volatilities. Applied Mathematical Finance, **2**, 73–88
4. Avellaneda, M., Parás, A. (1996): Managing the Volatility Risk of Portfolios of Derivative Securities: the Lagrangian Uncertain Volatility Model. Applied Mathematical Finance, **3**, 21–52
5. Avellaneda, M., Friedman, C., Holmes, R., Samperi, D. (1997): Calibrating Volatility Surfaces via Relative-entropy Minimization. Applied Mathematical Finance, **4**, 37-64
6. Avellaneda, M. (1998): Minimum-entropy Calibration of Asset-pricing Models. International Journal of Theoretical and Applied Finance, **1**, 447–472
7. Avellaneda, M., Buff, R., Friedman, C., Grandchamp, N., Kruk, L., Newman, J. (2001): Weighted Monte-Carlo: A New Technique for Calibrating Asset-Pricing Models. International Journal of Theoretical and Applied Finance, **4**, 91–120
8. Barraquand, J., Pudet, T. (1996): Pricing of American Path-dependent Contingent Claims. Mathematical Finance, **6**, 16–51
9. Ball, C. A., Roma, A. (1994): Stochastic Volatility Option Pricing. Journal of Financial and Quantitative Analysis, **29**, 589–607
10. Baxter, M., Rennie, A. (1996): Financial Calculus. Cambridge University Press, Cambridge
11. Bensoussan, A. (1984): On the Theory of Option Pricing. Acta Applicandae Mathematicae, **2**, 139–158
12. Black, F., Derman, E., Toy, W. (1990): A One-Factor Model of Interest Rates and its Application to Treasury Bond Options. Financial Analysts Journal, Jan/Feb 1990, 33–39
13. Black, F., Karasinski, P. (1991): Bond and Options Pricing when Short Rates are Lognormal. Financial Analysts Journal, July/August 1991, 52–59
14. Borodin, A. N., Salminen, P. (1996): Handbook of Brownian Motion—Facts and Formulae. Basel: Birkhäuser
15. Breeden, D. T., Litzenberger, R. H. (1978): Prices of State-contingent Claims Implicit in Option Prices. Journal of Business, **51**, 621–651
16. Buff, R. (1999): Algorithms for Nonlinear Models in Computational Finance and their Object-oriented Implementation. PhD Thesis, New York University, New York

17. Buff, R. (2000): Worst-case Scenarios for American Options. International Journal of Theoretical and Applied Finance, **3**, 25–58
18. Cheuk, T. H. F., Vorst, T. C. F. (1996): Complex Barrier Options. The Journal of Derivatives., **4**, 8–22
19. Conze, A., Viswanathan (1991): Path Dependent Options: The Case of Lookback Options. Journal of Finance, **46**, 1893–1907
20. Cormen, T. H., Leiserson, C. E., Rivest, R. L. (1990): Introduction to Algorithms. McGraw-Hill
21. Courtadon, G. (1982): A More Accurate Finite Difference Approximation for the Valuation of Options. Journal of Financial and Quantitative Analysis, **27**, 697–703
22. Cover, T. M., Thomas, J. A. (1991): Elements of Information Theory. Wiley, New York
23. Cox, J. C., Ingersoll Jr., J. E., Ross, S. A. (1985a): A Theory of the Term Structure of Interest Rates. Econometrica, **53**, 385–407
24. Cox, J. C., Ingersoll Jr., J. E., Ross, S. A. (1985b): An Intertemporal General Equilibrium Model of Asset Prices. Econometrica, **53**, 363–384
25. Cox, J. C., Rubinstein, M. (1985): Options Markets. Prentice Hall, New Jersey
26. Duffie, D. (1996): Dynamic Asset Pricing Theory, 2nd edition. Princeton University Press, New Jersey
27. Duffie, D., Kan, R. (1996): A Yield-factor Model of Interest Rates. Mathematical Finance, **6**, 379–406
28. Föllmer, H., Schweizer, M. (1991): Hedging of Contigent Claims under Incomplete Information. In: Davis, M. H. A., Elliott, R. J. (eds) Applied Stochastic Analysis, 389–414. Gordon and Breach, New York
29. Forsyth, P. A., Vetzal, K. R. (2001): Implicit solution of uncertain volatility/transaction cost option pricing models with discretely observed barriers. Applied Numercial Mathematics, **36**, 427–445
30. Fouque, J.-P., Papanicolaou, G., Sircar, K. R. (2000): Derivatives in Financial Markets with Stochastic Volatility. Cambridge University Press, Cambridge
31. Frey, R. (1997): Derivative Asset Analysis in Models with Level-Dependent and Stochastic Volatility. CWI Quarterly, **10**, 1–34
32. Geman, H., Yor, M. (1996): Pricing, and Hedging Double-barrier Options: a Probabilistic Approach. Mathematical Finance, **6**, 365–378
33. Geske, R., Shastri, K. (1985): Valuation by Approximation: A Comparison of Alternative Option Valuation Techniques. Journal of Financial and Quantitative Analysis, **20**, 45–71
34. Goldman, M. B., Sosin, H. B., Gatto, M. A. (1979): Path Dependent Options: Buy at the High, Sell at the Low. Journal of Finance, **34**, 1111-1127
35. Gozzi, F., Vargiolu, T. (1999): On the superreplication approach for European multiasset derivatives. International Conference on Mathematical Finance, 10th INFORMS Applied Probability Conference
36. Gozzi, F., Vargiolu, T. (2000): On the superreplication approach for European interest rate derivatives. Proceedings of the Ascona '99 Seminar on Stochastic Analysis, Random Fields and Applications
37. Heath, D., Jarrow, R., Morton, A. (1992): Bond Pricing, and the Term Structure of Interest Rates: a New Methodology for Contigent Claims Valuation. Econometrica, **60**, 77-105
38. Hofmann, N., Platen, E., Schweizer, M. (1992): Option Pricing under Incompleteness and Stochastic Volatility. Mathematical Finance, **2**, 153–187

39. Harrison, J. M., Kreps, D. M. (1979): Martingales and Arbitrage in Multiperiod Securities Markets. Journal of Economic Theory, **20**, 381–408
40. Harrison, J. M., Pliska, S. R. (1981): Martingales and Stochastic Integrals in the Theory of Continuous Trading. Stochastic Processes and Applications, **11**, 215–260
41. Heston, S. L. (1993): A Closed-form Solution for Options with Stochastic Volatility with Applications to Bond and Currency Options. Review of Financial Studies, **6**, 327–343
42. Ho, T. S. Y., Lee. S-B. (1986): Term Structure Movements and Pricing Interest Rate Contingent Claims. Journal of Finance, **41**, 1011–1029
43. Hull, J. C. (1993): Options, Futures and Other Derivative Securities, 2nd edition. Prentice Hall, New Jersey
44. Hull, J. C., White, A. (1987): The Pricing of Options on Assets with Stochastic Volatilities. Journal of Finance, **42**, 281–300
45. Hull, J. C., White, A. (1990): Pricing Interest Rate Derivative Securities. Review of Financial Studies, **3**, 573–592
46. Hull, J. C., White, A. (1990): Valuing Derivative Securities Using the Explicit Finite Difference Method. Journal of Financial and Quantitative Analysis, **25**, 87–100
47. Hull, J. C., White, A. (1993): Bond Option Pricing Based on a Model for the Evolution of Bond Prices. Advances in Futures and Options Research, **6**, 1–13
48. Hull, J. C., White, A. (1994): Numerical Procedures for Implementing Term Structure Models. Working paper, University of Toronto
49. Jackwerth, J. C., Rubinstein, M. (1996): Recovering Probability Distributions from Option Prices. Journal of Finance, **51**, 1611–1631
50. James, J., Webber, N. (2000): Interest Rate Modelling. Wiley, Chichester.
51. Jarrow, R. A. (1996): Modelling Fixed Income Securities and Interest Rate Options. McGraw-Hill
52. Jeanblanc-Picque, M., El Karoui, N., Viswanathan, R. (1991): Bounds for the Price of Options. In: I. Karatzas, I., Ocone, D. (eds) Applied Stochastic Analysis. Springer, New York
53. Johnson, H., Shanno, D. (1987): Option Pricing When the Variance is Changing. Journal of Financial and Quantitative Analysis, **22**, 143–151
54. Karatzas, I. (1988): On the Pricing of American Options. Applied Mathematics and Optimization, **17**, 37–60
55. Karatzas, I. (1989): Optimization Problems in the Theory of Continous Trading. SIAM Journal on Control and Optimization, **27**, 1221–1259
56. Klebaner, F. C. (1998): Introduction to Stochastic Calculus with Applications. Imperial College Press.
57. Kloeden, P. E., Platen, E. (1991): The Numerical Solution of Stochastic Differential Equations. Springer, New York
58. Kunitomo, N., Ikeda, M. (1992): Pricing Options with Curved Boundaries. Mathematical Finance, **2**, 275–272
59. Lagnado, R., Osher, S. (1997): A Technique for Calibrating Derivative Security Pricing Models: Numerical Solution of an Inverse Problem. Journal of Computational Finance, **1**, 13–25
60. Lamberton, D., Lapeyre, B. (1996): Introduction to Stochastic Calculus Applied to Finance. Chapman & Hall, London
61. LAPACK. http://gams.nist.com

62. Longstaff, F. A., Schwartz, E. S. (1998): Valuing American Options By Simulation: A Simple Least Squares Approach. Working paper, Anderson Graduate School of Management, University of California, Los Angeles
63. Lyden, S. (1996): Reference Check: A Bibliography of Exotic Options Models. The Journal of Derivatives, **4**, 79–91
64. Oksendal, B. (1998): Stochastic Differential Equations: An Introduction with Applications, 5th edition. Springer, New York
65. Parás, A. (1995): Non-linear Partial Differential Equations in Finance: a Study of Volatility Risk and Transaction Costs. PhD thesis, New York University, New York
66. Pirkner, C. D., Weigend, A. S, Zimmermann, H. (1999): Extracting Risk-Neutral Densities from Option Prices Using Mixture Binomial Trees. Proceedings of the 1999 IEEE/IAFE/INFORMS Conference on Computational Intelligence for Financial Engineering (CIFEr'99), 135–158
67. Press, H. W., Flannery, B. P., Teukolsky, S. A., Vetterling, W. T. (1993): Numerical Recipes in C, 2nd edition. Cambridge University Press, Cambridge
68. Roberts, C. O., Shortland, C. F. (1997): Pricing Barrier Options with Time-dependent Coefficients. Mathematical Finance, **7**, 93–93
69. Rubinstein, M. (1995): Implied Binomial Trees. Journal of Finance, **49**, 771–818
70. Rubinstein, M., Reiner, E. (1991): Breaking Down the Barriers. RISK, 4, 8
71. Schweizer, M. (1991): Option Hedging for Semimartingales. Stochastic Processes and Applications, **37**, 339–363
72. Scott, L. O. (1987): Option Pricing when the Variance Changes Randomly: Theory, Estimation and an Application. Journal of Financial and Quantitative Analysis, **22**, 419-438
73. Stein, E. M., Stein, J. C. (1991): Stock Price Distibutions with Stochastic Volatility: An Analytic Approach. Review of Financial Studies, **4**, 727–752
74. Stigum, M., Robinson, F. (1996): Money Market & Bond Calculations. Irwin Professional Publishing
75. Thomas, J. W. (1995): Numerical Partial Differential Equations: Finite Difference Methods. Springer, New York
76. Turnbull, S. M., Wakeman, L. M. (1991): A Quick Algorithm for Pricing European Average Options. Journal of Financial and Quantitative Analysis, **26**, 377–389
77. Wiggins, J. B. (1987): Option Values under Stochastic Volatility: Theory and Empirical Estimates. Journal of Financial Economics, **19**, 351–372
78. Wilmott, P., Dewynne, J., Howison, S. (1993): Option Pricing: Mathematical Models and Computation. Oxford Financial Press
79. Wilmott, P., Howison, S., Dewynne, J. (1995): The Mathematics of Financial Derivatives. Cambridge University Press
80. Vasicek, O. (1977): An Equilibrium Characterization of the Term Structure. Journal of Financial Economics, **5**, 177–188
81. Zhu, Y., Avellaneda, M. (1997): A Risk-Neutral Stochastic Volatility Model. Working paper, New York University
82. Zhu, C., Boyd, R. H., Lu, P., and Nocedal, J. (1994): L-BFGS-B: Fortran Subroutines for Large-scale Bound-constrained Optimization. Northwestern University, Department of Electrical Engineering

Index

ACALL[], 228
Accuracy, 229
accuracy, 218
addForward(), 160
addImplied(), 160
adversary, 125
affine, 23
afterRollback(), 178
afterRun(), 187, 189, 192
alpha, 209
AlwaysGrayScale, 231
american_put, 205
american_call, 205
American options, 77
apply(), 150
APUT[], 228
arbitrage-free price, 16
ask_price, 207
AssetPrice, 229

barrier, 51, 52, 61
base, 213, 215
beforeRollback(), 178
beforeRun(), 187, 189, 192, 194
beginIncrement(), 192
benchmarks, 30, 41
best worst-case, 80, 82
best worst-case price process, 83, 84, 86, 93, 103
bid_price, 207
Black, Derman and Toy, 24
Black-Karasinski, 23
Black-Scholes, 15, 17, 26–29, 37
blue, 216
bond price, 19
bootstrap, 203, 214
boundary conditions, 61
box, 212

Brownian motion, 15, 25, 34
burstiness of volatility, 29
butterfly spread, 57, 133
BUYER, 228
buyer, 216

calcWeights(), 180–182
calibration, 30, 36, 122
candidate set, 35, 62, 99, 123
carry, 209
cash bond, 20
certainUntil(), 162
checkpoint, 107
chooseOver(), 170
CIR, 23
claim, 203
claim(), 151, 169
closure, 64
combinatorial complexity, 77, 97
combinatorial phase, 78
complete, 25
computational complexity, 110
consolidate, 127, 131, 134
constantUntil(), 162
contingent claim, 15
continuation region, 50, 75, 77, 91, 94, 98
conventional, 127–129, 135
convergence, 117, 121
convex portfolio, 37
corridor of uncertainty, 91, 93, 97, 98, 107, 117
cost of money, 19
coupon, 207
Cox-Ingercoll-Ross, 23
Crank-Nicholson, 51, 71, 121, 133
createEngine(), 155, 157
createEvolution(), 155

238   Index

createInstance(), 176, 178, 195, 196
createOFSolver(), 175, 179
createSolver(), 192
createSpaceAxis(), 155, 156, 172
current(), 178
curve, 214
custom, 205, 207
cutoff rule, 91

day, 206, 208, 213
day(), 147
day_count, 213, 215
days_per_year, 208
DaysPerYear, 229
delta, 18
delta-hedging, 17, 58, 79
delta-neutral, 18
deterministic volatility, 15
Deviations, 229
digital, 205
directed acyclic graph, 107
discount, 209
discount bond, 19
discount curve, 31
discount process, 16, 34
discrete stopping time, 86
doBarriers(), 192
doBoundary(), 192
doIncrement(), 192
domestic, 209
domino effect, 105
doMonitor(), 192
dont_exercise, 206
DontExercise, 149, 169
doPayoff(), 192
doRollback(), 192
doRound(), 188
doTask(), 188, 190, 192
double-barrier options, 63, 70
down-and-out barrier, 52, 63, 70, 74
down_and_out, 205, 207
DownAndOut, 228
downBarrier(), 149
drift, 203, 210, 215
DRIFT[], 228
Duration, 229
duration, 124, 127
duration, 217

dynamic programming, 33, 47, 90, 123, 125, 139

early exercise, 77
early exercise boundary, 77, 91, 98, 125
early exercise combination, 78, 90, 93, 111
early exercise payoff, 98
early exercise region, 77
early exercise strategy, 80
ECALL[], 228
endIncrement(), 192
endureOver(), 170
entropy, 42
EPUT[], 228
European call, 57
European put, 57
european_call, 205
european_put, 205
Europeanization, 81
eval(), 196
evaluation space, 116
Exact, 229
exact, 155, 184, 218
exercise region, 91, 94, 98
exercisePayoff(), 148, 169
exotic volatility scenario, 123, 138
expected future payoff, 98
expected payoff, 77
Explicit, 229
explicit, 155, 173
explicit method, 50, 54
exponential complexity, 98
extended lattice signature, 127
extended Vasicek, 22
extension, 64
extension hierarchy, 65, 66, 106
external consistency, 50, 131

factor, 203, 206, 208
factor(), 147
faithfulness, 118
fast mean reversion, 29
filtered probability space, 15, 24, 34
finite difference lattice, 63
finite difference method, 40, 47, 54
firewall, 199
flux function, 40
following, 207

force_exercise, 206
forced exercise, 80
ForceExercise, 149, 169
forecasting problem, 123
foreign, 209
forward, 210, 211, 214
forward rate curve, 20
forward(), 161, 163, 165
fraction_of_day, 206, 208
free boundary, 86
frequency, 124, 135
full portfolio, 65

gamma, 18, 60
generate(), 149
getBarriers(), 147, 151, 155
getClaim(), 170, 188, 189
GetComplexityPlot[], 231
GetDependencyPlot[], 231
getEvents(), 147, 151
getEvGenerator(), 171
GetForceExPolicyPlot[], 232
getFwdRange(), 161
getLatticeInstance(), 188, 192
GetMayExPolicyPlot[], 232
getProcessParams(), 192, 193
GetTotal[], 232
GetVolatilityPlot[], 232
gnu_data, 214, 215
gnu_tics, 214, 215
go.bat, 8, 197
gradient vector, 78
green, 216

Heath-Jarrow-Morton, 20
hedge portfolio, 26, 38, 122
hedge ratio, 19
hedging, 17
Heston, 28
heuristic, 92, 98
high, 217
historical data, 30
HJM, 20
Ho and Lee, 23
Hull and White, 23, 26
hyperplane, 144

IgnoreLastSlice, 231
image, 203, 215

Implicit, 229
implicit, 155, 173, 212
implicit method, 51
implied, 210, 211
implied coefficients, 30
implied volatility, 30
implied(), 163
IMPLIED[], 228
incomplete, 25
initial, 210
inner loop, 66
instantaneous forward rate, 19
instantiation, 33
interest rate risk, 18
internal consistency, 49
irregular barriers, 93, 106
iterative improvement, 78
iterative refinement, 51

Johnson and Shanno, 27

kappa, 18
knock-out barrier, 51, 52, 61
knockout_payoff, 206
knockoutPayoff(), 148
Kullback-Leibler distance, 41

Lagrange multipliers, 42
largest up-and-out barrier, 64
last(), 178
lattice, 48
lattice, 203, 212
lattice boundary, 106
lattice instance, 49, 75, 78, 86, 93, 125, 129, 135
lattice node, 49, 59
lattice signature, 49, 63, 66, 73, 81, 85, 86, 93, 99, 103, 105
leg, 214
level-dependent, 24
linear, 205
lines, 216
liquidly traded assets, 30
local fixation, 82, 87, 102
local maximization, 84
local minimization, 84
local stability, 99
LONG, 228
long, 208

240   Index

long positions, 81
Low, 229
low, 155, 184, 217, 218
lower bound, 103
lower extension, 64, 69
lowest row maximum, 81

m_bIsBox, 173
m_bIsTrimmed, 173
m_bMonitor, 147
m_Bounds, 173
m_Buffer, 178
m_Cashflow, 147, 149
m_Claim, 151
m_CurveContainer, 184
m_Factor, 151, 154
m_FactorXlat, 187
m_gDiscount, 182
m_gDrift, 182
m_gFractionOfDay, 188
m_gFwd, 160
m_gFwd2, 160
m_gImp, 160
m_gMultiplier, 147
m_Gradient, 196
m_gRoot, 156
m_gVol, 182
m_ImageContainer, 184
m_Lambda, 196
m_nAccuracy, 184
m_nCertainUntil, 163
m_nConstantUntil, 163
m_nCurrent, 178
m_nCurTag, 194
m_nDay, 188
m_nDepth, 194
m_nDuration, 170
m_nFrequency, 170
m_nMaturity, 147
m_nMethod, 173
m_nPeriodicity, 170
m_nPosition, 168
m_nUnit, 160
m_pCalendar, 154
m_pCarry, 156
m_pCurCoInstance, 194
m_pDiscount, 156
m_pFDEngine, 184
m_pLattice, 184, 188

m_pMCEngine, 184
m_pModel, 172, 184, 187
m_pMu, 156
m_pOptimizer, 184, 195
m_Pos, 188
m_pPathSpace, 184
m_pPortfolio, 184, 187
m_Prep, 178
m_Price, 196
m_pScenario, 184, 188
m_pSolver, 192
m_pTime, 175
m_pVol, 156
m_Slot, 178
m_Space, 173
m_Spec, 160
m_Temp1, 178
m_Temp2, 178
m_Weight, 196
MakeDeltaFunc[], 231
MakeGammaFunc[], 231
MakePriceFunc[], 230
MakeScriptFunc[], 231
market index, 27
market price of volatility risk, 28, 29
martingale measure, 16, 21, 25
matchFactors(), 151, 187
MathLink, 227
maturity, 205
maturity(), 151
maximum-principle, 80
may_exercise, 206
MayExercise, 149, 169
mean, 209
mean reversion, 22
Method, 229
method, 212
minimize(), 196
minimum-entropy calibration, 41
mixed convexity, 43
model, 203, 208
modified, 207
monitor, 206
monitor(), 147, 149, 169
Monte-Carlo, 41
Mtg, 1
MtgClt, 143, 197, 203
MtgLib, 143

MtgMath, 144, 203, 227
MtgScript, 199, 203
MtgSvr, 143, 197, 203
multi-lattice, 48

von Neumann condition, 52
node instance, 49, 50, 91
non-parametric methods, 31, 41
nonlinearity, 43
nontradable asset, 25
numerical phase, 78

OpenFile[], 231
optimal hedge portfolio, 36, 38
optimizer, 203, 217
option premium, 18
Ornstein-Uhlenbeck, 22, 28
outer loop, 66
outlier, 120

parametric methods, 30
parse(), 194
partial differential equation, 48
partial portfolio, 49, 61, 63, 77, 90, 93, 105, 139
path-dependent, 35, 61
path_space, 203, 212
Payoff, 228
payoff, 205
payoff condition, 83
payoff from early exercise, 82
payoff(), 147, 150, 188
PDE, 48
penalty, 40
Periodicity, 229
periodicity, 124, 127
periodicity, 217
portfolio, 203, 208
PORTFOLIO[], 228
position, 208
positive interest rates, 23
positive volatility, 29
preceding, 207
preemptive solution, 107
prepare(), 156
price, 207
price band, 92
price vector, 39, 42
prior model, 41

prior volatility, 123, 127
Profile, 229
profile, 231
pseudo entropy function, 40

quantization problem, 123

random portfolio space, 115
random walk, 15
recursion, 63, 105
red, 216
refine(), 181, 192
refineExPolicy(), 169, 170, 188, 192
regularized signature, 167
reintroducing optionality, 83, 88
reintroducing uncertainty, 82
Repetitions, 229
repetitions, 217
replicating strategy, 16, 17, 25, 27
residual lattice instance, 82, 85, 102
residual worst-case liability, 39, 62
rho, 18
risk profile, 80
risk-diversification, 43
riskless asset, 15, 25
rollback, 144
root lattice instance, 167
root-lattice instance, 81, 83
rotate(), 178
round, 144
round-robin assignment, 131
round-robin index, 131
run(), 184, 186, 189
running time, 90, 107, 111, 118
RunScript[], 229

sample, 215
save, 213-216
scale, 213, 215
scaleDown(), 160
scaleUp(), 160
scenario, 33
scenario, 203, 216
scenario measure, 35
scenario volatility, 39, 40
SCENARIO[], 228
Scott, 27
Script[], 229
security price process, 15

seed, 213
selectVol(), 169, 170, 193
self-financing, 16, 25
sell-side, 79, 81
SELLER, 228
Seller, 169
seller, 216
Shape, 229
shape, 212
shock, 216, 218
shock front, 125
shock period, 124, 127, 129, 131, 134
shock volatility, 123
SHOCK[], 228
SHORT, 228
short, 208
short positions, 81
short rate, 20, 21, 24, 41
sigma, 209
signature, 49, 66, 73, 81, 85, 86, 93, 99, 103, 105
signature arithmetic, 86
signature tag, 167
single-barrier options, 70
singleton lattice instance, 97
singleton portfolio, 84
size, 213, 216
smallest up-and-out barrier, 64, 69
solve(), 179, 181
SOR, 77
space label, 49
spatial level, 52, 106
speedup, 92
Stein and Stein, 28
steps, 216
stochastic control, 33
stochastic volatility, 24
stress-tests, 115
sub-additive, 38, 86, 133
sub-hedging, 43
sub-lattice instance, 81, 83
subjective belief, 123
subjective prior beliefs, 39
successive over-relaxation, 77
super-hedging, 38, 43, 57
super-martingale, 37
system, 218
systematic risk, 26

tail period, 129
tArrayBounds, 173
tBondEquations, 194
tBondMath, 194
tBSModel, 153, 175, 182
tCashflow, 147, 149
tClaim, 146, 147, 169
tCustomClaim, 150
tDrift, 156, 158, 159, 162, 163
tEngine, 147, 155, 186, 193
tEntropyOpt, 196
termination, 77
tEvaluate, 184, 195
tExPolicyExpr, 194
tExPolicyExpression, 150
tExPolicyExpression, 151
tExpression, 194
tFDEngine, 166, 170, 185, 186, 188, 192, 193
tFileSource, 194
tGeoEngine, 157, 182, 186, 193
tGeoExplicit, 175, 180, 182
tGeoImplicit, 180, 182
tGeoSolver, 179, 182
tGeoSpaceAxis, 156, 175, 179
tHeap, 194
tHeap2, 194
theta, 18
theta, 209
tHJMGaussianModel, 153
tHJMTermStruct, 158, 159
time, 206, 208
time complexity, 71, 74, 135
time decay, 18
time label, 49
time_step, 212, 213
TimeStep, 229
tIncrement, 180, 190, 192
tJobServer, 194
tLattice, 172, 177
tLatticeInstance, 177, 178
tLinTermStruct, 162, 163
tMap, 160, 194
tMCEngine, 185
tMCEntropyOptInstance, 196
tMCOptInstance, 196
tMinimizer, 196
tModel, 152–155

tMultiArray, 194
tNetSource, 194
tNumericalExpr, 150, 151, 194
tOFEngine, 186, 190, 192, 193
tOFExplicit, 179, 181, 182
tOFImplicit, 179, 181
tOFSolver, 179–182, 190
tOptimizer, 195
tOptInstance, 195, 196
tParser, 194
tPathSpace, 184
tPortfolio, 151, 153, 187
tPosition, 168
tProcessParamsStub, 180, 182
tradable asset, 25
tree, 212
tridiagonal system, 51
tScanner, 194
tScenario, 167, 169, 171
tService, 194
tShockEngine, 157, 186, 193
tShockScenario, 157, 170
tShortRateTermStruct, 158, 159
tSignature, 176, 194
tSocket, 194
tSource, 194
tSpaceAxis, 173, 179
tSpec, 160
tSqTermStruct, 162, 163
tStdClaim, 150
tStepDrift, 165
tStepVol, 165
tStringSource, 194
tTermStruct, 162
tTermStruct, 158–160
tTimeAxis, 175, 179
tUSBondMath, 194
tUSTBond, 151
tVasicekModel, 153
tVol, 158, 159, 162, 163
tWorstCase, 157, 170

uncertain coefficients, 33
uncertain volatility, 33
underControl(), 169
uniform spacing, 52
up-and-out barrier, 52, 63, 70, 74
up_and_out, 205, 207
UpAndOut, 228
upBarrier(), 149
upper bound, 101
upper extension, 64
ust_bond, 207
UVM, 34

value/gradient pair, 49
vanilla options, 57
Vasicek, 22
vol, 203, 209, 211
VOL[], 227
volatility band, 110, 124, 127, 129, 138
volatility bounds, 52
volatility oscillation, 126, 135
volatility risk, 18
volatility shock date, 127
volatility smile, 33
volatility spike, 124, 133
volatility surface, 31, 41, 59
volatility-shock scenario, 34, 123, 125, 129
von Neumann condition, 52

weight, 213
Wiggins, 26
worst-case liability, 79, 80
worst-case pricing, 36, 61, 77
worst-case volatility scenario, 33, 35, 48, 57
worst-case volatility shock price, 132
worst_case, 216, 218
WORSTCASE[], 228

year, 213
yield, 214

Made in the USA
Middletown, DE
13 August 2023

36647600R00144